SpringerBriefs in Applied Sciences and Technology

SpringerBriefs present concise summaries of cutting-edge research and practical applications across a wide spectrum of fields. Featuring compact volumes of 50–125 pages, the series covers a range of content from professional to academic.

Typical publications can be:

- A timely report of state-of-the art methods
- An introduction to or a manual for the application of mathematical or computer techniques
- A bridge between new research results, as published in journal articles
- A snapshot of a hot or emerging topic
- An in-depth case study
- A presentation of core concepts that students must understand in order to make independent contributions

SpringerBriefs are characterized by fast, global electronic dissemination, standard publishing contracts, standardized manuscript preparation and formatting guidelines, and expedited production schedules.

On the one hand, **SpringerBriefs in Applied Sciences and Technology** are devoted to the publication of fundamentals and applications within the different classical engineering disciplines as well as in interdisciplinary fields that recently emerged between these areas. On the other hand, as the boundary separating fundamental research and applied technology is more and more dissolving, this series is particularly open to trans-disciplinary topics between fundamental science and engineering.

Indexed by EI-Compendex, SCOPUS and Springerlink.

More information about this series at http://www.springer.com/series/8884

Toan Dinh · Nam-Trung Nguyen
Dzung Viet Dao
Authors

Andreas Oechsner
Editor

Thermoelectrical Effect in SiC for High-Temperature MEMS Sensors

 Springer

Authors
Toan Dinh
Queensland Micro- and Nanotechnology
  Centre (QMNC)
Griffith University
Brisbane, QLD, Australia

Nam-Trung Nguyen
Queensland Micro- and Nanotechnology
  Centre (QMNC)
Griffith University
Brisbane, QLD, Australia

Dzung Viet Dao
School of Engineering and Built
  Environment
Griffith University
Southport, QLD, Australia

Editor
Andreas Oechsner
Esslingen University of Applied Sciences
Esslingen am Neckar, Germany

ISSN 2191-530X   ISSN 2191-5318   (electronic)
SpringerBriefs in Applied Sciences and Technology
ISBN 978-981-13-2570-0   ISBN 978-981-13-2571-7   (eBook)
https://doi.org/10.1007/978-981-13-2571-7

Library of Congress Control Number: 2018954596

This Springer imprint is published by the registered company Springer Nature Singapore Pte Ltd.
The registered company address is: 152 Beach Road, #21-01/04 Gateway East, Singapore 189721,
Singapore

Preface

There has been great interest and emerging demand for the development of aerospace sensing technology. This technology is used for sensing a wide range of systems working in harsh environments. The sensing applications in hostile conditions include, but not limited to, deep space exploration, combustion monitoring and observation of hypersonic aircraft. To maintain the safety and efficiency of instrumentation used in harsh environment industries, advanced health monitoring technologies are required to develop sensing networks which function reliably in harsh environments. However, the current technology has faced great challenges in providing sustainable solutions for the growth and stability of micro-/nanosystems. For example, the difficulty of locating sensing and electronic components in hostile environments has led to the utilisation of indirect measurement approaches which are expensive and inaccurate. These technologies have employed conventional materials such as silicon, which cannot withstand high-temperature and high-corrosive environments.

In addition, the resource sector, which includes mining, oil and gas and geothermal industries, is one of the driving forces of the global economy. The oil and gas delivery infrastructure is rapidly ageing. Internal corrosion and mechanical strain can cause the leakage in gas and oil pipelines, leading to catastrophic failures, death, injury and environmental impacts. Numerous sensor technologies have been used for monitoring and accident prevention purposes. However, most of the sensors are bulky or also based on materials like silicon (Si), which is not suitable for long-time operation in high-temperature environments found in pipelines, and geothermal and mining applications. Therefore, there have been new requirements for sensing technologies operating beyond the capability of Si sensing.

Silicon carbide (SiC) has been developed for sensing systems operating in harsh conditions including high-temperature, high-corrosion, high-voltage/power and high-frequency applications. Tremendous progress has been made in the development of microelectromechanical system (MEMS) sensors which employ SiC as a functional sensing material. These sensors have been commercialised for temperatures up to 300 °C. A number of recent studies have successfully demonstrated the application of SiC sensors for high temperatures above 500 °C, indicating a great potential for using this material for sensing in harsh environments.

This book focuses on the recent development of SiC thermal sensors employing high thermoelectrical effects in SiC at high temperatures. The thermoelectrical effect in SiC refers to a change in the electrical properties of SiC with temperature variation. This book will introduce the thermoelectrical effect in the form of the following physical effects: thermoresistive, thermoelectric, thermoelectronic and thermocapacitive effects in single and multiple SiC layers. The important characteristics of SiC materials and the fabrication processes for SiC MEMS are mentioned. The recent development of SiC sensing devices including temperature sensors, thermal flow sensors and convective inertial sensors is reviewed. The future prospects of SiC thermoelectrical sensing devices are discussed in this book.

Brisbane, Australia   Toan Dinh
Brisbane, Australia   Nam-Trung Nguyen
Southport, Australia   Dzung Viet Dao

Acknowledgements

The authors would like to acknowledge the support from Queensland Micro- and Nanotechnology Centre (QMNC) and School of Engineering and Built Environment at Griffith University, Australia. The authors are grateful for the contribution of the members of MEMS group at QMNC, including Dr. Hoang-Phuong Phan, Dr. Afzaal Qamar, Mr. Vivekananthan Balakrishnan, Mr. Tuan-Khoa Nguyen, Mr. Alan Iacopi, Ms. Leonie Hold, Mr. Glenn Walker and Mr. Abu Riduan Md Foisal, to the SiC MEMS project. The authors would like to thank the financial support from Australian Research Council grants LP150100153 and LP160101553. Toan Dinh is grateful the support from Griffith University & Simon Fraser University Collaborative Travel Grants Scheme 2017 and Australian Nanotechnology Network (ANN) Overseas Travel Fellowship 2018. We thank all authors and publishers for their graciousness to grant permission to reuse figures and tables in this book.

Contents

Chapter 1
Introduction to SiC and Thermoelectrical Properties

Abstract This chapter presents the general background on silicon carbide as a functional semiconductor for sensors operating in harsh environments. The fundamental stacking orders of different SiC polytypes with common growth methods and conditions are introduced, with a focus on cubic silicon carbide (3C-SiC) and hexagonal silicon carbide (e.g. 4H-SiC and 6H-SiC). This chapter also introduces the thermoelectrical effect in SiC with respect to sensing properties at high temperatures. The importance of SiC materials with a wide range of applications in harsh environments will be mentioned.

Keywords Silicon carbide · Thermoelectrical effect · MEMS Harsh environments

1.1 Background

There has been a high level of interest and emerging demand for the development of aerospace technology which is one of top-tier high-technology industries [1]. To maintain the safety and efficiency of instrumentation used in aerospace industries, advanced health monitoring technologies are required to develop sensing networks (e.g. Internet of Things) which can function in harsh environments [2–4]. However, these technologies have faced great challenges in providing sustainable solutions for the growth and stability of micro/nanosystems working in harsh conditions including deep space exploration, combustion monitoring and observation of hypersonic aircraft [5–7]. For example, the current electronic systems, including sensing and actuating instrumentation used in aerospace applications, have employed indirect measurement technologies which are expensive and inaccurate. This is due to the difficulty of locating sensing and electronic components in hostile environments [1, 8]. These components have deployed conventional materials such as silicon, which cannot withstand high-temperature and high-corrosive environments [8, 9]. Therefore, there have been new requirements for sensing technologies operating beyond

© The Author(s) 2018
T. Dinh et al., *Thermoelectrical Effect in SiC for High-Temperature MEMS Sensors*, SpringerBriefs in Applied Sciences and Technology, https://doi.org/10.1007/978-981-13-2571-7_1

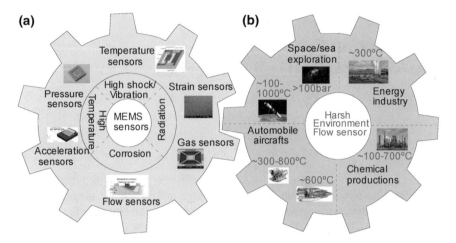

Fig. 1.1 Illustration of SiC MEMS sensors and harsh environments. **a** Type of SiC MEMS sensors with harsh environments including corrosion, high temperature, high shock/vibration and high radiation. **b** Applications of SiC MEMS sensors for high-temperature applications. Reprinted with permission from Ref. [5]

the capability of Si sensing [10]. Figure 1.1 shows the different harsh environments and the proposed SiC MEMS sensing technologies [5].

Wide band gap semiconductors including gallium nitride (GaN) and silicon carbide (SiC) have attracted great attention for a wide range of applications in harsh environments [11, 12]. Among them, SiC has certain advantages in terms of capability to be grown at a high quality and at low cost, as well as compatibility with conventional micro/nanomachining technologies, including microelectromechanical system (MEMS) and integrated circuit (IC) [13–15]. SiC has an extremely high sublimation temperature of 2830 °C due to the strong covalent bond between Si and C atoms. In addition, SiC offers excellent electrical stability at high temperatures because it has a wide band gap (e.g. 2.3 eV for 3C-SiC, 3.0 eV for 6H-SiC and 3.2 eV for 4H-SiC) which prevents the generation of intrinsic carriers at elevated temperatures [16, 17]. This property enables the development of SiC sensors and electronics working at high temperatures without the need for active cooling systems. Moreover, SiC has a high acoustic velocity of approximately 12,000 m/s, enabling the improvement of high bandwidth sensors [18, 19]. The excellent chemical inertness of SiC is also suitable for functioning as a sensing element, as well as a protective layer for devices operating in corrosive environments such as under sea water [20, 21].

1.2 Silicon Carbide

Silicon carbide (SiC) exhibits a one-dimensional polymorphism called polytypism, which exists in approximately 200 distinct crystalline polytypes [18, 22, 23]. These crystalline polytypes are differentiated by the stacking sequence of tetrahedrally bonded Si and C bilayers. The polytypes are categorised into three main basic crystallographies: cubic (C), hexagonal (H) and rhombohedral (R).

Cubic silicon carbide (3C-SiC) is a common type of crystallography used for sensing applications since it can be produced at a high quality. It is referred to as 3C-SiC or β-SiC with the number 3 referring to the number of layers. The stacking sequence of this 3C-SiC is shown in Fig. 1.2a. It is assumed that A, B and C denote the three layers; then, ABCABC…is its stacking sequence with cubic zinc blende crystallographic structure. To date, 3C-iC can be epitaxially grown on large-area Si wafers. However, one drawback of 3C-SiC grown on Si is the lattice mismatch, approximately 20% between 3C-SiC and Si, which leads to residual stress existing in SiC films [24].

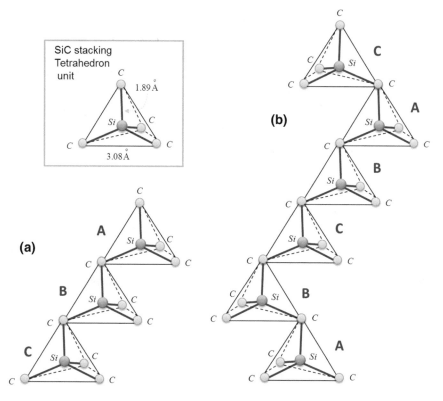

Fig. 1.2 Stacking order of **a** 3C-SiC and **b** 6H-SiC

Table 1.1 Physical characteristics of silicon carbide in comparison to silicon [13, 23]

Quantity	Si	3C-SiC	4H-SiC	6H-SiC
Band gap (eV)	1.1	2.4	3.2	3.0
Electron mobility (cm^2/Vs)	1400	800	1000	400
Hole mobility (cm^2/Vs)	471	40	115	101
Thermal conductivity (W/cm K)	1.3	3.6	4.9	4.9
Thermal expansion coefficient (ppm/K)	2.6	3.28	3.3	3.35
Melting point (K)	1690	3103	3103	3103
Young's modulus (GPa)	130–180	330–384	–	441–500
Density (g/cm^2)	2.33	3.21	3.21	3.21

If the stacking sequence of a bi-layer is ABAB..., the hexagonal type is the symmetry referred to as 2H-SiC. 4H-SiC consists of an equal number of cubic and hexagonal bonds, while 6H-SiC composes of two-thirds cubic bonds and one-third hexagonal bonds. Figure 1.2b shows the stacking sequence of 6H-SiC. Both 4H-SiC and 6H-SiC are grouped as α-Si and has been commercially available in bulk wafer form. Due to its superior properties and large band gap, SiC has attracted recent research for MEMS devices operating in harsh environments. Table 1.1 shows the physical characteristics of three common SiC polytypes in comparison with the conventional MEMS material Si.

1.3 Growth of SiC

Silicon wafer processing is the most mature and central semiconductor technology. The low-cost and high-quality Si wafers are typically employed to grow 3C-SiC films [25, 26]. As Si naturally has a cubic crystalline structure, SiC films grown on Si inherit the crystal structure to form a cubic structure. Therefore, 3C-SiC on Si has become an attractive platform technology for electronics and sensors. Generally, SiC can be grown using different techniques such as radiofrequency (rf), magnetron sputtering, hot-wire chemical vapour deposition (CVD) and low-pressure chemical vapour deposition (LPCVD). These methods can grow single-crystalline, nanocrystalline (nc-SiC) and amorphous (a-SiC) structures, depending upon the growth conditions and substrates for growing SiC. In chemical vapour deposition processes, single-crystalline SiC grown on Si substrates requires a relatively high temperature of 1000–1200 °C with two containing precursors of Si and C, such as silane SiH$_4$ and methane CH$_4$. The formation of SiC films depends on a few parameters including growth temperature, pressure and gas flow rate. Conventionally, the growth of single-crystalline SiC requires a high temperature of above 1000 °C, while a temperature of 400–800 °C can be used to grow a-SiC and nc-SiC. In comparison to other growth technologies, LPCVD involves high deposition temperatures to decompose the precursor sources and also enhances the chemical order of SiC structures and doping efficiency.

4H-SiC and 6H-SiC cannot be grown on cubic silicon substrates as silicon has a naturally cubic structure. These polytypes have a hexagonal structure and typically can be grown on the same substrate. The growth of 4H- and 6H-SiC requires very high growth temperature ranging from 1800 to 2400 °C [13]. Therefore, the cost of 4H- and 6H-SiC wafers is very high compared to 3C-SiC grown on Si substrates. The details of growth processes are presented in Chap. 4. To avoid the leakage current to the substrate, 4H-SiC wafers are typically formed with p–n junction. Consequently, this requires to selectively etching of n-type or p-type substrates to form the functional 4H-SiC structures. In addition to the fabrication difficulty, other complex processing procedures such as Ohmic contact to 4H-SiC also present great challenges for the development of sensing systems on chips [27, 28].

1.4 Thermoelectrical Properties

The intrinsic SiC materials typically have a low conductivity. To be employed for sensing applications, SiC materials are doped with n- or p-type conductivity for ease of electrical measurement [29, 30]. The common range of resistivity for thermal sensors and MEMS devices is from 0.01 to 10^4 Ω cm.

Thermoelectrical effect refers to the change of the electrical properties of SiC to temperature variation. Figure 1.3 shows the category of the thermoelectrical effect, which includes four main effects of temperature on the electrical properties of SiC, namely thermoresistive, thermoelectric, thermocapacitive and thermoelectronic effects. Among them, the thermoresistive and thermoelectric effects are typically measured for a single layer of SiC, while the thermocapacitive and thermoelectronic effects are found in multiple layers of different doping types of SiC or between a metal layer and SiC layer.

The working principle of the thermoresistive effect in SiC is described as follows. When the temperature increases, the impurities are ionised and contribute to the conduction of SiC. Therefore, the conductivity of SiC increases with increasing

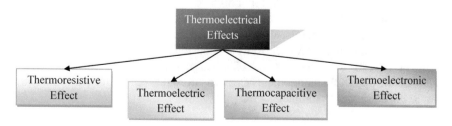

Fig. 1.3 Category of thermoelectrical effects

temperature, corresponding to a decrease of the electrical resistivity [31]. However, at high doping levels, all impurities can be ionised at room temperature, leading to a decrease of the electrical conductivity, or increase of the electrical resistivity. The increase of the electrical resistance with increasing temperature is governed by the scattering effect. It is important to note that the thermoelectrical properties of SiC are stable at high temperatures (e.g. 600 °C) due to its large band gap preventing the generation of intrinsic carriers. Advancement in doping of impurities to SiC micro/nanostructures can also offer controllable thermoelectrical properties at high temperatures, enabling the development of thermal sensors in harsh environments [32, 33].

The thermoelectronic concept is employed to the electronic structures of multiple layers of SiC such as diodes and transistors [34, 35]. To evaluate the electrical properties of SiC thermoelectronic devices for sensing applications, the current–voltage (*I–V*) characteristic of SiC devices is typically employed. For example, the electrical resistance of SiC devices is used to define the change of electrical properties of SiC p–n junctions (e.g. *I–V* characteristic of the interface between an n-type SiC layer and a p-type SiC layer). At a constant applied current, the voltage changes that occur with the changing temperatures are deployed to evaluate the temperature sensitivity of the system. This voltage change is typically linear with changing temperatures in 4H-SiC structures. The sensitivity found for the 4H-SiC and 6H-SiC p–n junctions and 4H-SiC Schottky diodes commonly ranges from 1 to 5 mV/K [34, 36]. However, the thermoelectrical properties of 3C-SiC p–n junctions have not yet been fully understood. This is probably due to the low quality of 3C-SiC p–n junctions.

Thermocapacitive principle involves change of the capacitance with the temperature variation. In general, the increase of carrier concentration with increasing temperature leads to an increase of the capacitance [37]. The thermocapacitive in SiC has been intensively studied. However, the advantage of this sensing effect does not outperform other thermoelectrical sensing effects, which results in a less successful demonstration of this effect for sensing applications at high temperatures. The thermoelectric effect in semiconductors refers to the generation of electric potential difference between two points when a temperature difference between them is applied [38]. To achieve a high efficiency of generating electric power, the Seebeck coefficient should be high and the thermal conductivity should be small. However, SiC holds a relatively high thermal conductivity and resistivity, leading to the small figure of merit for collecting electric power.

I–V characteristics in a single SiC film/layer can be converted to the electrical resistance by Ohm's law $R = V/I$. The temperature sensitivity of the SiC film is measured using the temperature coefficient of resistance (TCR), which is defined by the resistance change ($\Delta R/R$) over the temperature variation (ΔT). The typical range of TCR for SiC materials is 2000–5000 ppm/K for highly doped materials, 4000–16000 ppm/K for amorphous SiC, and up to 20,000 ppm/K for unintentionally n-type doped single-crystalline 3C-SiC [39].

1.5   High-Temperature SiC MEMS Sensors

MEMS sensors employing Si have been well developed with numerous successful applications. The main advantage of MEMS technology is the miniaturisation capability for devices, enabling the development of multiple sensors on a single chip [40]. The small size allows MEMS sensors with low cost, high sensitivity and fast response. MEMS sensors employing conventional Si material have been developed with a high level of maturity, and there are a number of commercialised sensors available. The microfabrication for MEMS devices has been well established with advanced etching techniques. The conventional surface and bulk-micromachining technologies for Si MEMS are also able to fabricate SiC MEMS sensors as both SiC and Si share common chemical and physical properties.

In addition to the comparability of SiC to MEMS fabrications, there has been a great need for the development of MEMS sensors operating in harsh environments including high temperatures and highly corrosive conditions. SiC MEMS is a suitable choice for these applications as SiC has a large band gap and excellent chemical inertness. The stability of SiC allows the development of highly sensitive SiC MEMS sensors at high temperatures. These SiC sensors can replace the current indirect measurement technologies for high temperatures, owing to the excellent properties of SiC for sensing and its capability to locate SiC sensors directly into systems. This enables real-time measurements of structural health.

Integration of SiC sensors into such systems at high temperatures requires all supporting devices/structures to also work at high temperatures. Therefore, high-temperature contacts and wiring would also need to be addressed for full integration of SiC into the systems working in high temperatures.

References

1. D.G. Senesky, B. Jamshidi, K.B. Cheng, A.P. Pisano, Harsh environment silicon carbide sensors for health and performance monitoring of aerospace systems: a review. IEEE Sens. J. **9**, 1472–1478 (2009)
2. J.A. Erkoyuncu, R. Roy, E. Shehab, P. Wardle, Uncertainty challenges in service cost estimation for product-service systems in the aerospace and defence industries, in *Proceedings of the 19th CIRP Design Conference—Competitive Design* (2009)
3. T.Q. Trung, N.E. Lee, Flexible and stretchable physical sensor integrated platforms for wearable human-activity monitoring and personal healthcare. *Advanced Materials* (2016)
4. H. Sohn, C.R. Farrar, F.M. Hemez, D.D. Shunk, D.W. Stinemates, B.R. Nadler et al., *A Review of Structural Health Monitoring Literature: 1996–2001* (Los Alamos National Laboratory, USA, 2003)
5. V. Balakrishnan, H.-P. Phan, T. Dinh, D.V. Dao, N.-T. Nguyen, Thermal flow sensors for harsh environments. Sensors **17**, 2061 (2017)
6. Y. Wang, Y. Jia, Q. Chen, Y. Wang, A passive wireless temperature sensor for harsh environment applications. Sensors **8**, 7982–7995 (2008)
7. H. Kairm, D. Delfin, M.A.I. Shuvo, L.A. Chavez, C.R. Garcia, J.H. Barton et al., Concept and model of a metamaterial-based passive wireless temperature sensor for harsh environment applications. IEEE Sens. J. **15**, 1445–1452 (2015)

8. L. Chen, M. Mehregany, A silicon carbide capacitive pressure sensor for high temperature and harsh environment applications, in *Solid-State Sensors, Actuators and Microsystems Conference, 2007. TRANSDUCERS 2007. International* (2007), pp. 2597–2600

9. K.S. Szajda, C.G. Sodini, H.F. Bowman, A low noise, high resolution silicon temperature sensor. IEEE J. Solid-State Circuits **31**, 1308–1313 (1996)

10. R.G. Azevedo, D.G. Jones, A.V. Jog, B. Jamshidi, D.R. Myers, L. Chen et al., A SiC MEMS resonant strain sensor for harsh environment applications. IEEE Sens. J. **7**, 568–576 (2007)

11. M.E. Levinshtein, S.L. Rumyantsev, M.S. Shur, *Properties of Advanced Semiconductor Materials: GaN, AlN, InN, BN, SiC, SiGe* (Wiley, London, 2001)

12. R.F. Davis, Thin films and devices of diamond, silicon carbide and gallium nitride. Phys. B **185**, 1–15 (1993)

13. J. Casady, R.W. Johnson, Status of silicon carbide (SiC) as a wide-bandgap semiconductor for high-temperature applications: a review. Solid-State Electron **39**, 1409–1422 (1996)

14. M. Mehregany, C.A. Zorman, N. Rajan, C.H. Wu, Silicon carbide MEMS for harsh environments. Proc. IEEE **86**, 1594–1609 (1998)

15. X. She, A.Q. Huang, Ó. Lucía, B. Ozpineci, Review of silicon carbide power devices and their applications. IEEE Trans. Ind. Electron. **64**, 8193–8205 (2017)

16. L. Wang, A. Iacopi, S. Dimitrijev, G. Walker, A. Fernandes, L. Hold et al., Misorientation dependent epilayer tilting and stress distribution in heteroepitaxially grown silicon carbide on silicon (111) substrate. Thin Solid Films **564**, 39–44 (2014)

17. G.L. Harris, *Properties of silicon carbide* (IET, 1995)

18. D. Feldman, J.H. Parker Jr., W. Choyke, L. Patrick, Phonon dispersion curves by raman scattering in SiC, Polytypes 3 C, 4 H, 6 H, 1 5 R, and 2 1 R. Phys. Rev. **173**, 787 (1968)

19. G.N. Morscher, A.L. Gyekenyesi, The velocity and attenuation of acoustic emission waves in SiC/SiC composites loaded in tension. Compos. Sci. Technol. **62**, 1171–1180 (2002)

20. K.N. Lee, R.A. Miller, Oxidation behavior of muilite-coated SiC and SiC/SiC composites under thermal cycling between room temperature and 1200°–1400 °C. J. Am. Ceram. Soc. **79**, 620–626 (1996)

21. L. Shi, C. Sun, P. Gao, F. Zhou, W. Liu, Mechanical properties and wear and corrosion resistance of electrodeposited Ni–Co/SiC nanocomposite coating. Appl. Surf. Sci. **252**, 3591–3599 (2006)

22. D. Barrett, R. Campbell, Electron mobility measurements in SiC polytypes. J. Appl. Phys. **38**, 53–55 (1967)

23. M. Mehregany, C.A. Zorman, SiC MEMS: opportunities and challenges for applications in harsh environments. Thin Solid Films **355**, 518–524 (1999)

24. H. Mukaida, H. Okumura, J. Lee, H. Daimon, E. Sakuma, S. Misawa et al., Raman scattering of SiC: estimation of the internal stress in 3C-SiC on Si. J. Appl. Phys. **62**, 254–257 (1987)

25. L. Wang, S. Dimitrijev, J. Han, A. Iacopi, L. Hold, P. Tanner et al., Growth of 3C–SiC on 150-mm Si (100) substrates by alternating supply epitaxy at 1000 C. Thin Solid Films **519**, 6443–6446 (2011)

26. L. Wang, S. Dimitrijev, J. Han, P. Tanner, A. Iacopi, L. Hold, Demonstration of p-type 3C–SiC grown on 150 mm Si (1 0 0) substrates by atomic-layer epitaxy at 1000 °C. J. Cryst. Growth **329**, 67–70 (2011)

27. F. Roccaforte, F. La Via, V. Raineri, Ohmic contacts to SiC. Int. J. High Speed Electron. Syst. **15**, 781–820 (2005)

28. Z. Wang, W. Liu, C. Wang, Recent progress in ohmic contacts to silicon carbide for high-temperature applications. J. Electron. Mater. **45**, 267–284 (2016)

29. K. Nishi, A. Ikeda, D. Marui, H. Ikenoue, T. Asano, n-and p-Type Doping of 4H-SiC by Wet-Chemical Laser Processing, in *Materials Science Forum* (2014), pp. 645–648

30. K. Eto, H. Suo, T. Kato, H. Okumura, Growth of P-type 4H–SiC single crystals by physical vapor transport using aluminum and nitrogen co-doping. J. Cryst. Growth **470**, 154–158 (2017)

31. S.M. Sze, K.K. Ng, *Physics of Semiconductor Devices* (Wiley, 2006)

32. P. Wellmann, S. Bushevoy, R. Weingärtner, Evaluation of n-type doping of 4H-SiC and n-/p-type doping of 6H-SiC using absorption measurements. Mater. Sci. Eng., B **80**, 352–356 (2001)

33. A. Kovalevskii, A. Dolbik, S. Voitekh, Effect of doping on the temperature coefficient of resistance of polysilicon films. Russ. Microlectron. **36**, 153–158 (2007)
34. S. Rao, G. Pangallo, F.G. Della Corte, 4H-SiC pin diode as highly linear temperature sensor. IEEE Trans. Electron Devices **63**, 414–418 (2016)
35. S. Rao, G. Pangallo, F. Pezzimenti, F.G. Della Corte, High-performance temperature sensor based on 4H-SiC schottky diodes. IEEE Electron Device Lett. **36**, 720–722 (2015)
36. S.B. Hou, P.E. Hellström, C.M. Zetterling, M. Östling, 4H-SiC PIN diode as high temperature multifunction sensor, in *Materials Science Forum* (2017), pp. 630–633
37. S. Zhao, G. Lioliou, A. Barnett, Temperature dependence of commercial 4H-SiC UV Schottky photodiodes for X-ray detection and spectroscopy. Nucl. Instrum. Methods Phys. Res., Sect. A **859**, 76–82 (2017)
38. S. Fukuda, T. Kato, Y. Okamoto, H. Nakatsugawa, H. Kitagawa, S. Yamaguchi, Thermoelectric properties of single-crystalline SiC and dense sintered SiC for self-cooling devices. Jpn. J. Appl. Phys. **50**, 031301 (2011)
39. T. Dinh, H.-P. Phan, A. Qamar, P. Woodfield, N.-T. Nguyen, D.V. Dao, Thermoresistive effect for advanced thermal sensors: fundamentals, design considerations, and applications. J. Microelectromech. Syst. (2017)
40. J.W. Gardner, V.K. Varadan, O.O. Awadelkarim, *Microsensors, MEMS, and Smart Devices*, vol. 1 (Wiley Online Library, 2001)

Chapter 2
Fundamentals of Thermoelectrical Effect in SiC

Abstract This chapter presents the fundamentals of thermoresistive effect in different SiC morphologies including single-crystalline cubic SiC, polycrystalline and amorphous SiC. The thermocapacitive and thermoelectric effects are also summarised. In addition, recent advances in the characterisation of the thermoelectrical effect in SiC single and double layers with a heterostructure will be discussed. Other aspects of temperature effect on the electrical properties of SiC are mentioned.

Keywords Thermoelectrical effect · Thermoresistive effect · Thermoelectric
Thermoelectronic · Thermocapacitive

2.1 Thermoresistive Effect

Thermoresistive effect in silicon carbide (SiC) refers to its electrical resistance change with temperature. Generally, the electrical resistance R of SiC films can be described as follows:

$$R = \rho \frac{l}{wt} \tag{2.1}$$

where ρ is the resistivity of the material, respectively; l, w and t are the length, width and thickness of the SiC film, respectively. From Eq. (2.1), a small relative resistance change depends upon the resistivity and dimension changes [1, 2]:

$$\frac{\Delta R}{R} = \frac{\Delta \rho}{\rho} + \frac{\Delta l}{l} - \frac{\Delta w}{w} - \frac{\Delta t}{t} \tag{2.2}$$

where $\Delta\rho/\rho$ is the relative resistivity change; $\Delta l/l$, $\Delta w/w$ and $\Delta t/t$ are the changes of length, width and thickness of SiC films. The change of dimension due to temperature change ΔT can be expressed as: $\Delta l/l = \Delta w/w = \Delta t/t = \alpha \Delta T$, where α is the thermal expansion coefficient (TEC) of SiC and $\Delta T = T - T_o$ is the temperature variation;

© The Author(s) 2018
T. Dinh et al., *Thermoelectrical Effect in SiC for High-Temperature MEMS Sensors*, SpringerBriefs in Applied Sciences and Technology,
https://doi.org/10.1007/978-981-13-2571-7_2

11

Table 2.1 Thermoresistive in single-crystalline SiC in comparison with metals [4, 5, 7]

Materials	Resistivity $(10^{-8}\ \Omega m)$	TCR (ppm/K)	TCE, α (ppm/K)	Thermal expansion contribution (%)
Nickel (Ni)	6.84	6,810	12.7	0.19
Iron (Fe)	9.71	6,510	10.6	0.16
Tungsten (W)	5.5	4,600	4.3	0.09
Copper (Cu)	1.67	4,300	16.8	0.39
Aluminium (Al)	2.69	4,200	25.5	0.61
Silver (Ag)	1.63	4,100	18.8	0.46
Platinum (Pt)	10.6	3,920	8.9	0.23
Gold (Au)	2.3	3,900	14.3	0.37
Molybdenum (Mo)	5.57	4,820	4.8	0.10
Silicon carbide	1.4×10^5	−5,500	4.0	0.07

T and T_0 are the absolute temperature and reference temperature (typically room temperature). Therefore, Eq. (2.2) can be written in the following form [3]:

$$\frac{\Delta R}{R} = \frac{\Delta \rho}{\rho} - \alpha \Delta T \tag{2.3}$$

The temperature coefficient of resistance (TCR) can be deduced from Eq. (2.3) as follows [3]:

$$TCR = \frac{\Delta R}{R} \frac{1}{\Delta T} = \frac{\Delta \rho}{\rho} \frac{1}{\Delta T} - \alpha \tag{2.4}$$

Conventional thermal sensing materials including metals have a TCR of 3,900–6,800 ppm/K while the TEC value is less than 30 ppm/K [4, 5], corresponding to less than 0.61% contribution to the TCR (see Table 2.1). For simplicity in calculation, the impact of TEC to the calculated TCR can be neglected. Moreover, the contribution of geometric effects to the TCR of semiconductor is less than 5 ppm/K, while the TCR value of semiconductors used for thermal sensing applications is typically from several thousands to several ten thousands ppm/K, depending upon temperature ranges, doping levels and morphologies. For example, the geometry of silicon (TCR = −6,000 ppm/K [6]) contributes to 2.6 ppm/K, corresponding to an error of 0.04%. Therefore, the geometric effects are neglected when the materials are considered for temperature sensing purposes. Table 2.1 shows the TCR of single-crystalline SiC in comparison to conventional metals used for thermal sensing applications.

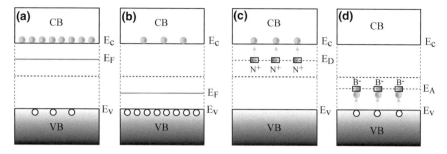

Fig. 2.1 Band gap diagrams. **a** Femi level in n-type semiconductors. **b** Femi level in p-type semi-conductors. **c** Donor energy in n-type semiconductors. **d** Acceptor energy in p-type semiconductors

2.1.1 Physical Parameters and Definitions in Semiconductors

The energy band gap is the smallest energy gap between the conduction band (CB) and the valence band (VB) [8, 9]. If the energy levels E_c and E_v mark the bottom and top of the conduction band and the valance band, the band gap is defined as $E_g = E_c - E_v$. The Femi level is a hypothetical energy level of an electron, having 50% probability of being occupied at thermodynamic equilibrium, which lies between E_c and E_v. There are two main semiconductor types, namely intrinsic and extrinsic semiconductors.

An intrinsic semiconductor is a pure semiconductor crystal, which has equal electron and hole concentration ($n = p$) and no impurities. By introducing impurities into a pure crystal, extrinsic semiconductors with different concentrations of electron and hole ($n \sim p$) are formed. Extrinsic semiconductors include n-type ($n > p$) semiconductor with a Femi level (EF) closer to CB, and p-type ($p > n$) semiconductor with a Femi level closer to VB. Figure 2.1a, b illustrates the band gap and Femi levels in n-type and p-type semiconductors.

There are two types of impurities, namely donor (n-type) and acceptor (p-type). Donor (E_D) and acceptor (E_A) energies are the energy levels, which are neutral and negative, respectively, if filled with an electron. The energy required to excite an electron to jump to CB, or a hole to jump to VB, is an activation energy. The activation (ionisation) energy of donors is $E_d = E_C - E_D$ and that of acceptors is $E_a = E_A - E_D$. Figure 2.1c, d illustrates the donor and acceptor energy levels in n-type and p-type semiconductors.

2.1.2 Single-Crystalline SiC

It is expected that the resistivity of single-crystalline SiC depends upon the carrier concentration and the carrier mobility as follows [8, 9]:

$$\frac{1}{\rho} = \sigma = qn\mu_e + qp\mu_h \tag{2.5}$$

where σ the conductivity and q is the electron charge, respectively; n and p are the electron and hole concentrations, correspondingly; μ_e and μ_h denote the corresponding electron and hole mobility.

In extrinsic region, corresponding to the low-temperature regime (typically lower than room temperature), carriers are generated by thermal energy and then excited to jump to the conduction band (CB). The carrier concentration is expressed as follows [7, 10]:

$$n \sim T^\alpha \exp\left(\frac{E_d}{kT}\right) \tag{2.6}$$

where E is the activation/ionisation energy, α (e.g. 3/2 for n-Si) is a constant, and k is the Boltzmann constant. In addition, lattice scattering causes a reduction of mobility as follows [7, 11, 12]:

$$\mu \sim T^\beta \tag{2.7}$$

where β is a constant (e.g. $\beta = 3/2$ for extrinsic silicon). From Eqs. (2.5), (2.6) and (2.7), the dependence of resistivity on temperature is written as [7]:

$$\rho \sim T^{\alpha-\beta} \exp\left(\frac{E_d}{kT}\right) \tag{2.8}$$

In various semiconductors, α and β have the same value and the resistivity is exponentially dependent on temperature $\rho \sim \exp(E_d/kT)$. Therefore, the dependence of resistivity on temperature T is described as:

$$\ln(\rho) = \frac{E_d}{k} \times \frac{1}{T} \tag{2.9}$$

Figure 2.2 shows the decrease in electrical resistivity in the extrinsic region with a slope of E_d/kT. In the metal-like region (higher temperature range), the mobility till decreases while the carrier concentration is stable, leading to an increase of the electrical resistivity. At high temperatures (intrinsic region), material bonds are ruptured, leading to a significant increase of carrier concentration. In this region, the temperature dependence of resistivity is described as $\rho \sim \rho_o \exp(E_g/2kT)$, indicating a decrease of the electrical resistivity with a slope of $E_g/2kT$. Figure 2.3 shows the temperature sensing mechanism for single-crystalline SiC with the generation of carriers.

For temperature sensing purposes, the dependence of the electrical resistance on temperature is typically used as follows [13, 14]:

Fig. 2.2 Thermoresistive effect in single-crystalline semiconductors

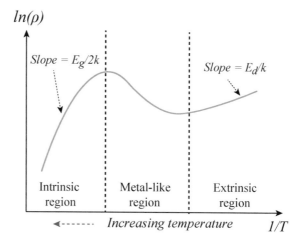

Fig. 2.3 Thermoresistive mechanism in single-crystalline materials

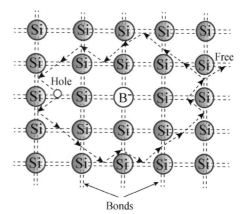

$$R = A \exp\left[B\left(\frac{1}{T} - \frac{1}{T_o}\right)\right] \tag{2.10}$$

where A and B are a constant and the thermal index, respectively. The thermal index B value is employed to evaluate the thermal sensing effect in thermistor-type temperature sensors. B is calculated as $E_g/2kT$ for intrinsic semiconductors and E_d/kT for doped semiconductors. The relationship between B and TCR is as follows: $B = -TCR \times T^2$ [14, 15].

2.1.3 Polycrystalline SiC

Apart from single-crystalline materials, a polycrystalline material is an assembly of crystallites and the boundaries between them. The defects and boundaries between

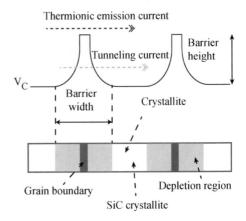

Fig. 2.4 Thermoresistive mechanism in polycrystalline materials

crystallites play dominant role in determination of electrical resistivity of the poly-crystalline materials. As such, the boundaries between the crystallites trap their surrounding carriers, forming a potential barrier between the crystallites [16]. This barrier impedes the movement of carriers through the barrier under electric field, leading to reduction of the carrier mobility. Therefore, the total resistance of poly-crystalline materials includes the crystallite resistance and the boundary resistance ($R = R_{\text{Boundary}} + R_{\text{Crystallite}}$) with the dominance of the latter [17]. An approximation is to consider the resistance of polycrystalline materials as the only resistance of the grain-boundary region.

The electrical resistance of boundaries depends upon the temperature change through two sensing mechanisms, including thermionic emission and field emission (tunnelling) (Fig. 2.4). Thermionic emission refers the transport of carriers over the potential barrier, owing to their high energy. Other carriers with low energy will pass through the barrier by quantum tunnelling mechanism [18, 19]. If the barrier is narrow and high enough, the tunnelling current can be comparable and even higher than thermionic emission current. However, when the polycrystalline material is doped at high levels, the barrier is low; hence, the thermionic emission mechanism is dominant. The temperature-dependent resistance of polycrystalline materials can be described as follows [20, 21]:

$$R = A \exp\left(\frac{\phi}{kT}\right) \tag{2.11}$$

It is important to note that in polycrystalline metals, the TCR can be tuned from a positive value to negative value via controlling their grain size. This is attributed to the fact that the overall TCR of polycrystalline metals is the combination of the positive TCR of their crystallites and negative TCR of their boundaries. The number of boundaries increases with decreasing grain size [22]. Therefore, the TCR value can turn to a negative value if the TCR of bulk metals is smaller than temperature coefficient of bulk mean free path.

2.1.4 Amorphous SiC

Amorphous semiconductors have an energy level between localised and extended states, named mobility edge [11, 23]. These states have a density of state (DOS), which can be a constant, parabolic or exponent function of temperature. The temperature-dependent resistance of amorphous materials is determined as follows:

$$\sigma = \sigma_o \exp\left[-\left(\frac{E_a}{kT}\right)^\beta\right] \tag{2.12}$$

where E_a denotes the activation energy and k is the Boltzmann constant. According to variable-range hopping theory, DOS functions as a constant, β value is ¼ [24, 25]. At low temperatures, DOS has a parabolic distribution, resulting in a β value of ½. At temperatures high enough, electron transitions between localised states and localised band tail play an important role on the temperature-dependent conductivity of amorphous materials with $\beta = 1$, showing an activation of electrons over the mobility edge [26].

2.2 Thermoelectronic Effects

Thermoelectronic effect refers to the change of electrical properties of semiconductor junctions with temperature variation. The typical temperature sensing elements are p–n junction diodes, Schottky diodes and transistors. This section mainly focuses on the impact of temperature on to the current–voltage (I–V) of p–n junctions.

Considering a p–n junction of semiconductors, the band gap diagram is shown in Fig. 2.5 [9]. Without electric field, the built-in potential E_o drives a balance between the diffusion current (n–p) and drift current (p–n), resulting in zero current between two electrodes. The diffusion current is $J_o = B \exp[-eV_o/kT]$. In the forward bias, the potential barrier reduces to $q(E_0 - E)$, leading to the diffusion of electrons from n-side to p-side with a diffusion current of $J_o = B \exp[-e(V_o - V)/kT]$. The forward current then can be expressed as follows [9, 27]:

$$J = J_o\left[\exp\left(\frac{eV}{kT}\right) - 1\right] \tag{2.13}$$

When temperature increases, the carriers are excited and collected at the electrodes under the electric field (Fig. 2.5b). In the reverse bias of $V = -V_r$, the diffusion current is very small as $J_o = B \exp[-e(V_o + V_r)/kT]$. Figure 2.6a shows the band diagram of the p–n junction under an application of a reverse bias. When the temperature increases, the thermal energy generates electron–hole pairs in the depletion layer, as well as the n-type and p-type materials. The electrons and holes are separated

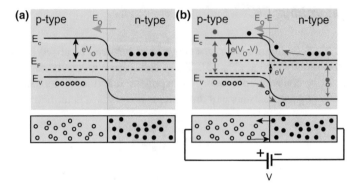

Fig. 2.5 Sensing mechanism of p–n thermojunctions under forward bias. **a** Energy band diagram of p–n junction with a built-in potential. **b** Carriers are generated at elevated temperatures and biased by a forward electric field

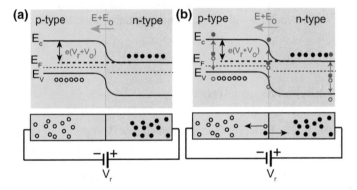

Fig. 2.6 Sensing mechanism of p–n thermojunctions under reverse bias. **a** Energy band diagram of p–n junction with a built-in potential and applied reverse bias. **b** Carriers are generated at elevated temperatures and biased by a reverse electric field

by the electric field and collected at the electrodes. This creates a reverse current J (Fig. 2.6b).

2.3 Thermocapacitive Effect

A *p–n* junction or a Schottky diode can be considered as a parallel capacitor as it has positive and negative charges separated by a distance W at the depletion region. This depletion layer width W is defined as [28]:

$$W = \left[\frac{2\varepsilon(N_a + N_d)(V_o - V)}{e N_a N_d} \right]^{1/2}$$

(2.14)

The depletion layer capacitance is determined as:

$$C_{dep} = \frac{\varepsilon A}{W} = \frac{A}{(V_o - V)^{1/2}} \left[\frac{e\varepsilon(N_a N_d)}{2(N_a + N_d)} \right]^{1/2} \qquad (2.15)$$

where ε is the semiconductor permittivity; N_a and N_d are the concentration of acceptors and donors, respectively. It is evident that the junction capacitance depends upon the applied voltage V. For example, under a reverse bias, the width of the depletion layer increases, leading to a decrease in the junction capacitance. Thermocapacitance effect refers to the change of the junction capacitance due to temperature change. It is expected that the junction capacitance increases with increasing temperature, considering the increase of carrier concentration under applied thermal energy. The temperature-dependent permittivity could also play a role on the variation of the junction capacitance. The thermocapacitive effect was used to develop flow sensors [29]. The dependence of the dielectric constant was employed to control the capacitance; hence, the capacitance change measured the change of the flow velocity.

2.4 Thermoelectric Effect

Thermoelectric effect refers to the generation of electric potential difference dV across a semiconductor or metal under application of temperature difference dT [30, 31]. This is also called Seebeck effect $S = dV/dT$. Figure 2.7 shows the Seebeck effect in metals/semiconductors [9]. When a metal/semiconductor is heated at one end and cooled at the other, the electrons in the hot region will have more energy and diffuse towards the cold end. This results in more positive ions at the hot end and more electrons at cold end. The voltage different between two ends is the Seebeck voltage. Table 2.2 shows the Seebeck coefficient of silicon carbide and several common metals.

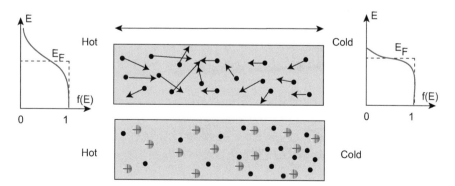

Fig. 2.7 Seebeck effect in a conductor. The temperature gradient results in a potential difference

Table 2.2 Seebeck coefficient of metals [9, 32–35]

Materials	Seebeck coefficient @ 0 °C (μV/K)	Seebeck coefficient @ 25 °C (μV/K)	E_F (eV)
Al	-1.6	-1.8	11.6
Au	$+1.79$	$+1.94$	5.5
Cu	$+1.70$	$+1.84$	7.0
K	–	-12.5	2.0
Li	$+14$		4.7
Mg	-1.3		7.1
Na	–	-5	3.1
Pd	-9	-9.99	–
Pt	-4.45	-5.28	–
SiC	–	10–100	–

2.5 Recent Advances in Characterisation of Thermoelectrical Effects in SiC at High Temperatures

Silicon carbide (SiC) is of high interest for MEMS thermal sensors operating in high temperatures, owing to its large band gap and superior sensing properties [36–38]. In single-crystalline SiC, the thermoresistive effect has been examined through the temperature effect on carrier concentration and carrier mobility. For example, the hole mobility in p-type 3C-SiC is driven by acoustic phonon scattering (above 300 K) and ionised impurity scattering (below 250 K), following the rule $\mu \sim T^{-\beta}$, where β value is from 1.2 to 1.4. The carrier concentration has been investigated to increase with increasing temperature $n \sim \exp(-E_d/kT)$, where E_d was found in a wide range for different polytypes. Therefore, the TCR value was typically found to be negative. However, in highly doped 3C-SiC, TCR is positive (400–7,200 ppm/K), which could expect to the dominance of scattering effect at high temperatures [39–41]. Recently, the quality of SiC films has been significantly improved and few advanced processes have been employed to lower the cost of SiC wafers. Therefore, tremendous progress has been made for the characterisation of SiC thermoelectrical effects. The following sections will provide the recent advancement in the development of SiC for high-temperature sensors.

2.5.1 Experimental Set-up for Characterisation of Thermoelectrical Effect

Figure 2.8 shows the conventional set-up for characterisation of the thermoelectrical effect at high temperatures [11, 38]. For example, a heater is packed in an enclosed

Fig. 2.8 Experimental set-up for characterisation of the thermoelectrical effect at high temperatures. Reprinted with permission from Ref. [11, 38]

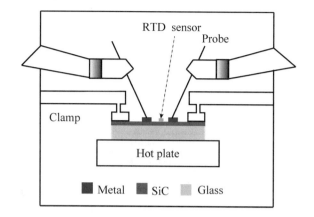

chamber and employed to heat the system up to a high temperature of 600 °C. A SiC sample is fixed on top of the heater by clamps, and probes push against the electrodes to electrically connect the SiC sensor with the electrical measurement unit. Wire boding can alternatively be used to make electrical connections. To have an accuracy measurement, a reference temperature sensor is placed on top of the SiC sensor. This set-up method has advantages over the utilisation of ovens because it does not require supporting components withstanding in high temperatures.

2.5.2 Thermoresistive Effect in Single Layer of SiC

Amorphous SiC [11]

A fused quartz substrate (1 inch square and 2 mm thick) was employed to grown SiC films using a low-pressure chemical vapour deposition (LPCVD) at 650 °C [11]. A single precursor of monomethylsilane (H_3SiCH_3) was used to deposit the SiC film in a pressure of 0.6 Pa at 9.5 sccm (standard cubic centimetres per minute) for ten hours. The thickness of the a-SiC film was determined to be 95 ± 3 nm using a spectroscopic ellipsometry method. The transmittance spectrum of a-SiC on glass in a range of 200–1100 nm was recorded as shown in Fig. 2.9. The plot of $(\alpha h v)^{1/2}$ versus hv (α, h and v are the optical absorption coefficient, Planck's constant and the frequency, respectively) points out an optical energy gap of 3.5 eV for the a-SiC film. The absorption coefficient was defined to be in the order of 10^5 cm^{-1} (the top right inset in Fig. 2.9). The semiconductor type of the SiC film was found as an unintentionally n-type doped conductivity employing the polarity of the hot probe voltage.

Figure 2.10a shows the decrease of the electrical resistance of a-SiC with increasing temperatures. The dependence of the electrical resistance R of a-SiC is described by the following relationship [25, 26]:

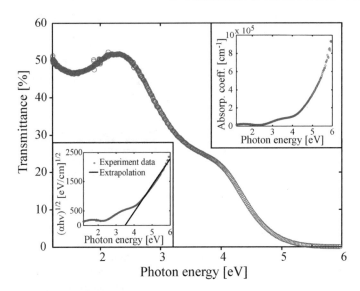

Fig. 2.9 Transmittance spectrum of a-SiC on glass. The bottom left inset reveals the optical band gap of SiC, and the top right inset shows the absorption coefficient of the a-SiC on glass. Reprinted with permission from Ref. [11]

$$R \sim \exp\left(\frac{-E_a}{k_B T}\right) \qquad (2.16)$$

where E_a and k_B are the activation energy and Boltzmann constant, respectively. Consequently, the conductivity change vs the temperature is plotted in Fig. 2.10b using $-\ln(\sigma/\sigma_o) = E_a(k_B T)^{-1} - b$; where σ and σ_o are the conductivity of the a-SiC film at temperature T and reference temperature T_o, respectively; b is a constant. Two well-fitted lineshapes were found for temperature ranges of below and above 450 K, corresponding to two activation energy thresholds of 150 and 250 meV, respectively [11, 42]. The increase of the activation energy with increasing temperature indicates the domination of the ionised donor states at elevated temperatures. The temperature coefficient of resistance (TCR) is approximated as the following relationship [11]:

$$TCR = \frac{\Delta R}{R}\frac{1}{\Delta T} = \frac{\exp\left[E_a\left(\frac{1}{k_B T} - \frac{1}{k_B T_o}\right)\right] - 1}{T - T_o} \qquad (2.17)$$

where $\Delta R/R$ is the resistance change and $T - T_o$ is the temperature difference. Figure 2.10c shows the resistance change and TCR plot for the full range of temperature. For example, $\Delta R/R$ reaches 96% at 580 K while the TCR value ranges from −16,000 to −4,000 ppm/K, indicating a high sensitivity of a-SiC films compared to conventional temperature sensing materials [13–15, 43, 44].

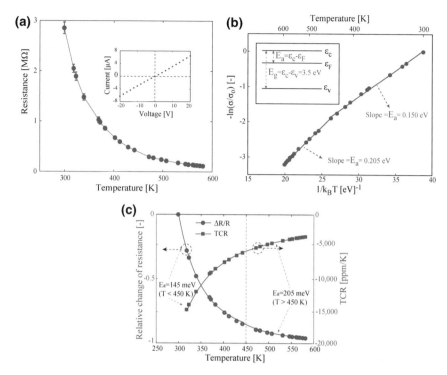

Fig. 2.10 Thermoresistance of a-SiC. **a** Resistance change with temperature variation. The inset shows the current–voltage characteristics of a-SiC measured at room temperature of 25 °C. **b** Arrhenius plot of a-SiC thermoresistance. The inset shows the sketch of energy levels in the unintentionally n-type doped a-SiC. **c** Relative resistance change of the a-SiC film and its temperature coefficient of resistance TCR. Reprinted with permission from Ref. [11]

Single-crystalline highly doped n-type 3C-SiC on Si [45]

A low-pressure chemical vapour deposition (LPCVD) at 1250 °C was employed to grown SiC nanothin films with a thickness of 600 nm on 625-μm-thick p-type Si(100) wafers [45]. The large SiC on Si wafer was formed with a diameter of 150 mm. Two precursor gases including silane SiH_4 and propene C_3H_6 were deployed to grow the SiC films, while ammonia NH_3 was utilised to in situ dope SiC films with n-type impurity at a high doping level of 10^{19} cm^{-1}.

Figure 2.11a shows the current–voltage characteristic of the SiC temperature sensor, indicating a good Ohmic contact formed with the electrodes [45]. Figure 2.11b shows the reliance of the SiC resistance on the applied temperature. The resistance increased from 684 Ω at the 25 °C to 700 Ω to above 120 °C [45]. It is evident that the resistance change is linear with increasing temperature and described by the following relationship [3]:

$$R = R_o + \alpha(T - T_o) \tag{2.18}$$

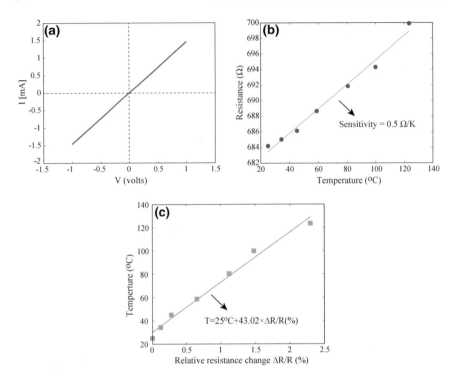

Fig. 2.11 Thermoresistive effect in highly doped SiC nanofilms. **a** Current–voltage characteristic of SiC at room temperature. **b** The temperature dependence of the SiC resistance. **c** Temperature of the SiC device as a function of its resistance change. Reprinted with permission from Ref. [45]

where R_o and R are the resistance at the room temperature T_o (25 °C) and elevated temperature T. From the fitting line, a resistance change of 0.5 Ω/K was calculated for the SiC thermoresistors. For temperature measurement of SiC devices, the dependence of temperature on the relative resistance change is deduced as $T = 25C + 43.02 \times \Delta R/R$ (%) (Fig. 2.11c) [45]. The temperature coefficient of resistance of highly doped SiC was found to be relatively low at approximately 250 ppm/K. Further studies should be conducted to investigate the thermoresistive characteristics of highly doped n-type SiC at high temperatures (e.g. above 600 °C).

Single-crystalline p-type 3C-SiC [7]

p-type 3C-SiC thin films with a thickness of 280 nm were grown on a (100) Si substrate, employing a LPCVD process at a temperature of 1273 K. Two precursors of silane (SiH$_4$) and propylene (C$_3$H$_6$) were utilised, and trimethylaluminium [(CH3)3Al, TMAl] was employed to form p-type SiC semiconductor. Figure 2.12a depicts the X-ray diffraction (XRD) spectrum of the as-grown SiC films. An approximately 0.8° was measured for the full width at haft maximum (FWHM) (Fig. 2.12b). In addition, the transmission electron microscope (TEM) image in Fig. 2.12c indicates the stacking faults existing in the interface between SiC and Si. Figure 2.12d

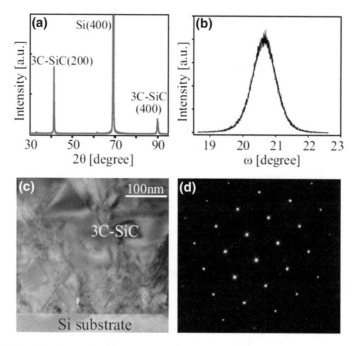

Fig. 2.12 Material characterisation of 3C-SiC: **a** the X-ray diffraction (XRD) graph of p-type 3C-SiC grown on a (100) Si substrate; **b** the rocking curve of 3C-SiC; **c** the transmission electron microscopy (TEM) image of 3C-SiC; **d** the selected area electron diffraction SAED image of 3C-SiC. Reproduced with permission from Ref. [46]. Copyright [2014], AIP Publishing LLC.

shows the selected area electron diffraction (SAED), indicating single-crystalline characteristics of the SiC films.

Figure 2.13a shows the resistance change with temperature up to 600 K. For example, the resistance decreased up to 80% at 600 K. Figure 2.13b shows the corresponding temperature coefficient of resistance (TCR) which is estimated from −2400 ppm/K to −5500 ppm/K [7]. This TCR value is comparable to that of conventional materials for thermal sensing such as platinum (3920 ppm/K). The thermoresistive effect in p-type 3C-SiC is attributed to the increase of the hole concentration (n) with increasing temperature. As a result, the TCR value is relative large around room temperature and decreases with increasing temperature as a result of the full ionisation of free holes and the decrease in the hole mobility. The dependence of the hole concentration is assumed as [8, 47]:

$$n \sim T^{3/2} \exp\left(\frac{E_a}{kT}\right) \tag{2.19}$$

where k and E_a are the Boltzmann constant and the activation energy, respectively. The decrease of the hole mobility with increasing temperature is approximated as [8, 12, 48]

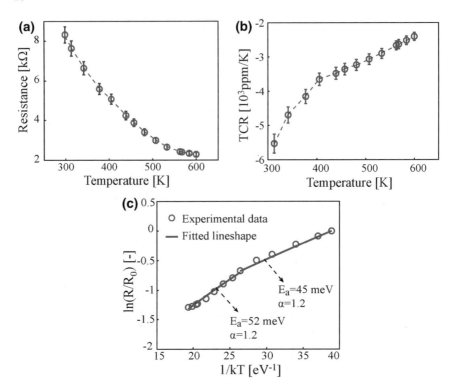

Fig. 2.13 Thermoresistive properties of p-type 3C-SiC: **a** resistance of SiC with temperature variation; **b** temperature coefficient of resistance (TCR) of p-type 3C-SiC; **c** Arrhenius plot of p-type SiC. Reprinted with permission from Ref. [7]

$$\mu \sim T^{-\alpha} \tag{2.20}$$

where α is an experimental constant. The resistivity of the SiC sensor is defined from the following relationship:

$$\rho = \frac{1}{q\mu n} \sim T^{\alpha - 3/2} \exp\left(\frac{E_a}{kT}\right) \tag{2.21}$$

Therefore, the resistance change of SiC can be described as follows:

$$\ln\left(\frac{R}{R_0}\right) = \left(\alpha - \frac{3}{2}\right) \ln\left(\frac{T}{T_0}\right) + E_a\left(\frac{1}{kT} - \frac{1}{kT_0}\right) \tag{2.22}$$

where R_0 and R are the SiC resistance at the reference temperatures T_0 and T, respectively. Figure 2.13c shows the Arrhenius plot of the p-3C-SiC with fitted curves. Two activation energies of 45 and 52 meV were calculated for the temperature range of below and above 450 K, respectively. In addition, the hole mobility constant was

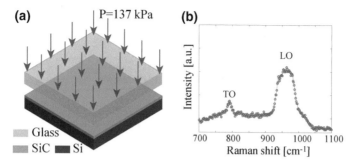

Fig. 2.14 a Unintentionally n-type doped SiC film transferred to a glass substrate by anodic bonding. **b** Raman spectrum of SiC on glass. Reprinted with permission from Ref. [38]

extracted to be approximately 1.2. Due to the large carrier concentration of 5×10^{18} cm^{-3}, the impurities are partly ionised at room temperature and the Fermi level is closer to the valence band [49]. Understanding the thermoresistive properties of 3C-SiC at high temperatures will open the door for the development of highly sensitive thermal sensors at high temperatures.

Unintentionally doped 3C-SiC [38]

Low-pressure chemical vapour deposition (LPCVD) at a temperature of 1000 °C was employed to grown single-crystalline cubic silicon carbide films on a (111) Si substrate. Two alternating precursors of SiH_4 and C_3H_6 were utilised to supply carbon atoms in the growth process. The unintentionally n-type doped SiC on Si films were formed during the standard growth processes at Queensland Microtechnology Facility (QMF), Griffith University, Australia. This is attributed to the occurrence of nitrogen residual gas.

To date, high-quality 3C-SiC has been grown on a Si substrate to produce SiC wafers with large area and low cost. However, large leakage current to the silicon substrate was found at elevated temperatures [48], which prevents the development of SiC sensors working at high temperatures. Therefore, SiC films were transferred onto an insulation to avoid the current leakage. As such, an ionic bonding technique was deployed to transfer the SiC films on to glass substrate (Borofloat 33, University Wafers) at a pressure of 137 kPa and bias voltage of 1000 V. The bonding between SiC and glass can withstand a maximum stress of 20 MPa. The Si layer was removed by mechanical polishing and wet etching approaches. Figure 2.14a shows the formation of SiC-on-glass platform using the anodic bonding method. The SiC material was confirmed with the Raman technique (Fig. 2.14b), showing two dominant peaks at the wavenumbers of 797 and 965 cm^{-1}, corresponding to the transverse optic (TO) mode and longitudinal optic (LO) mode of 3C-SiC material [50, 51], respectively. Other characteristics of SiC films were also characterised [52, 53].

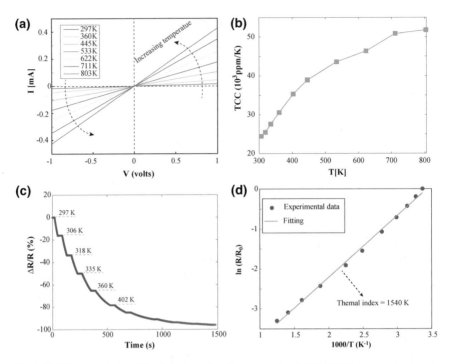

Fig. 2.15 Thermoresistive effect in unintentionally n-type doped SiC at high temperatures up to 800 K. **a** Current–voltage properties at elevated temperatures. **b** Temperature coefficient of conductivity (TCC). **c** Response of SiC to temperature variation. The inset illustrates the graph of the SiC response to the high temperature range of above 445 K. **d** Arrhenius plot for SiC temperature sensor. Reprinted with permission from Ref. [38]

Figure 2.15a depicts the linear current–voltage properties of SiC at elevated temperatures. At a constant applied voltage of 1 V, the measured current increased from 16.2 to 427.35 μA when the temperature increased from 300 to 800 K. This indicates the thermally activated conduction of SiC at elevated temperatures. The temperature coefficient of conductivity can be defined as $S_\sigma = \Delta\sigma/\sigma_o \times 1/\Delta T = (I - I_o)/I_o \times 1/\Delta T$. For instance, the TCC value increased 2.65 times from 19,550 to 51,780 ppm/K for the temperature range from 297 to above 800 K (Fig. 2.15b). This TCC value is larger than that of various conventional materials utilised for thermal sensors such as metals and silicon [3, 4, 54]. The high TCC shows the high interest of using the unintentionally doped SiC films for thermal sensors. Figure 2.15c shows the real-time response of the SiC sensor to the temperature change.

The electron concentration and electron mobility are two main factors contributed to the resistance change of SiC with temperature variation. The electron mobility is defined as $\mu = (1/\mu_l + 1/\mu_i)^{-1}$ where μ_l represents the lattice scattering and μ_i denotes the scattering of ionised impurities. μ_i is neglected at a high temperature of above 300 K due to the fact that SiC is unintentionally doped. The decrease of the electron mobility due to acoustic phonon scattering plays a crucial role for the temperature dependence of the SiC conductivity, which is described as $\mu_l \sim T^{-\alpha}$ [55, 56], where α is the mobility constant. The electron concentration is assumed to exponentially increase with increasing temperature $n \sim T \exp(-E_a/kT)$, where E_a and k are the activation energy and the Boltzmann constant, respectively. As 3C-SiC has crystal symmetry, the intervalley scattering is insignificant; hence, the temperature dependence is neglected in comparison with the temperature dependence of the electron concentration [57]. Therefore, the conductivity of the unintentionally n-type doped SiC material is defined as follows:

$$\sigma \sim \exp\left(\frac{-E_a}{kT}\right) \tag{2.23}$$

This can be represented in the following form:

$$R = R_0 \exp\left[B\left(\frac{1}{T} - \frac{1}{T_0}\right)\right] \tag{2.24}$$

where R_0 and R are the SiC resistance at the reference temperature T_0 and the elevated temperature T, respectively. $B = E_a/k$ denotes the thermal index of the SiC temperature detector. The resistance change vs temperature is determined as:

$$\ln\left(\frac{R}{R_0}\right) = \frac{B}{T} - A \tag{2.25}$$

where $A = B/T_0$ is a constant. Therefore, the thermal index was found to be 1540 K as shown in Fig. 2.15d. This thermal index value is comparable with other high-performance thermistors reported in the literature [58–60].

2.5.3 Thermoelectrical Effect in Multi-layers of SiC

Single-crystalline SiC on Si

It is well known that high-quality single-crystalline cubic silicon carbide (n-3C-SiC) can be grown on large-area Si substrates to make a SiC/Si platform [61–65]. To understand the transport mechanism in this platform, both current flow in the SiC layer (I_{SiC}) and the current flow from the SiC layer to the Si substrate (I_{Si}) were measured [66–69]. Figure 2.16a shows the measured currents in the SiC/Si platform, indicating the increase of both SiC current and the leakage current. Figure 2.16b shows the normalised current of SiC flowing in this platform, which is defined as $\Delta I_{SiC} = I_{SiC}/(I_{SiC} + I_{Si}) \times 100\%$, as well as the normalised leakage current flowing to the Si substrate $\Delta I_{Si} = I_{Si}/(I_{SiC} + I_{Si}) \times 100\%$. It is evident that the current leakage flowing in a Si substrate increases dramatically with increasing temperature, and ΔI_{Si} reaches almost 50% of the total current. This indicates the unsuitability of the SiC/Si platform for temperature sensors and power electronic devices which

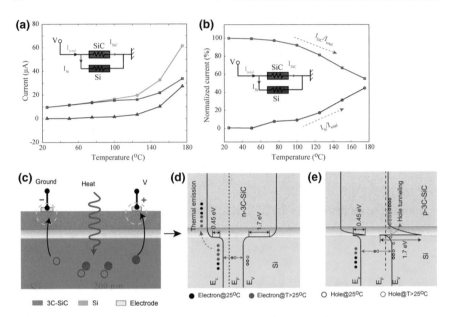

Fig. 2.16 Temperature effect on the SiC/Si platform. **a** Current flows in the SiC/Si platform; **b** normalised current in the SiC/Si platform; **c** SiC/Si platform showing the generation of carriers with increasing temperature; **d** conduction mechanism of n-type 3C-SiC/Si; **e** conduction mechanism of p-type 3C-SiC/Si. Reprinted with permission from Ref. [45]

require a typically working temperature of above 100 °C. Figures 2.16c, d show the temperature sensing mechanism for the failure of the SiC/Si heterostructures. The band offsets of 0.45 and 1.7 eV between 3C-SiC and Si were unable to prevent the motion of carriers through the SiC/Si heterostructure [69–71]. The lattice mismatch between SiC and Si leads to the existence of the stacking faults and boundaries in the SiC/Si heterostructure. The number of carriers in the Si substrate increases with increasing temperatures, resulting in the thermal emission of the electrons over the energy barrier (0.45 eV for n-type 3C-SiC). This leads to an electrical current flowing to the Si substrate. Figure 2.16e illustrates the transport mechanism for p-type SiC/Si, where the tunnelling current through the barrier (1.7 eV) is attributed as the main contribution to the current leakage at elevated temperature [45].

The 3C-SiC films that are grown on a Si substrate will induce an interface between 3C-SiC and Si (Figs. 2.17a). The corresponding energy band diagrams of the n-3C-SiC on Si and p-3C-SiC on Si are presented in Fig. 2.17b. A novel fabrication strategy was proposed to completely eliminate the leakage current as shown in Fig. 2.17c. The concept of the new platform is to remove the Si substrate and replace with SiO_2. This will eliminate the source of the carriers generated by thermal energy, as well as to introduce a high barrier to avoid the leakage current. Figure 2.17d shows the band diagrams of the n-3C-SiC/glass and p-3C-SiC/glass structures, indicating a very high barrier created by the SiO_2 layer.

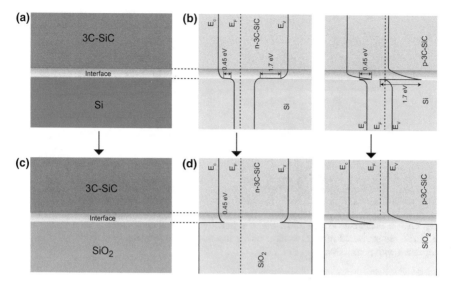

Fig. 2.17 **a** Single-crystalline 3C-SiC grown on Si with **b** its corresponding band energy structure for n-type and p-type SiC on Si. **c** A novel platform of the 3C-SiC on glass with **d** its corresponding band energy structure. Reprinted with permission from Ref. [45]

2.6 4H-SiC p–n Junctions

4H-SiC has been proven to show tremendous prospects for high power electronics working in harsh environments including high-temperature, high-corrosion and high-voltage applications. The integration of temperature sensing element in a single 4H-SiC power electronic chip with its capability of operating at high temperatures up to 600 °C is of great interest for a wide range of applications such as gas turbines and geothermal power plants [72–77]. The integration ability at a critical location on the device such as hot spots will enable the development of real-time monitoring systems. This improves the safety and efficiency of such systems as it will monitor and predict the failure in real time. The key considerations for design of sensing systems working in harsh environment are not only the high sensitivity but also the ease of integration of the sensing modules into a single chip with simple circuitry. A few designs of temperature modules have been successfully demonstrated, including diodes and field-effect transistors. Among these configurations, diodes are the simplest design and easiest in implementation.

4H-SiC could be employed for high-temperature sensing application as it has highest energy band gap of 3.2 eV in comparison to other polytypes of SiC. Recent research has demonstrated the current flow through the 4H-SiC p–n junction as a sensitive temperature sensing element. Figure 2.18a shows the structure of a 4H-SiC p–n junction temperature sensor working up to 600 °C with a Ni contact for n-type and a Ni/Ti/Al contact for p-type 4H-SiC. The active area of the device is 150 μm ×

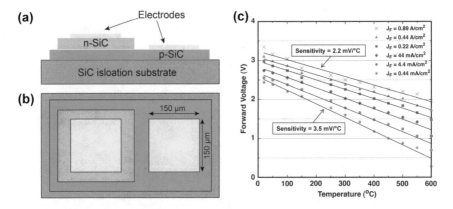

Fig. 2.18 **a** Schematic sketch of 4H-SiC p–n junction temperature sensors; **b** SEM image of the sensor; **c** the measured forward voltage of the junction under different applied current densities. Reprinted with permission from Ref. [78]

150 μm or $A = 2.25 \times 10^{-4}$ cm^2 (Fig. 2.18b). The current density is $J = I/A$, where I is the measured current that is described as follows [8, 9]:

$$I = I_o\left(e^{\frac{qV}{nkT}} - 1\right) \qquad (2.26)$$

where q and k are the electric charge and Boltzmann constant, respectively; n denotes the ideality factor which is typically found to be 2 for elevated temperatures with the dominant contribution of the recombination current; I_o is the saturation current. Under an application of a constant current I for the whole temperature range, the voltage dependence of on temperature is described as follows with assumption of $qV \gg nkT$:

$$V = \frac{2kT}{q}\ln\left(\frac{I}{I_o}\right) = \frac{2kT}{q}\ln\left(\frac{I\tau_e}{qWA}\right) - \frac{kT}{q}\ln(N_cN_v) + \frac{E_g}{q} \qquad (2.27)$$

where N_c and N_v are the density of the conduction band state and valence band state, respectively; τ_e denotes the carrier life time. Therefore, the sensitivity of p–n junction temperature sensors is theoretically defined as [78]:

$$\frac{dV}{dT} = \frac{2k}{q}\ln\left(\frac{I\tau_e}{qWA}\right) - 7.67 \text{ mV/K} \qquad (2.28)$$

Experimentally, the sensitivity of the p–n junction temperature sensor is shown in Fig. 2.18c. The maximum sensitivity of 3.5 mV/K was measured for the current density of 0.44 mA/cm^2. The increase of the current density leads to the decrease of the sensitivity.

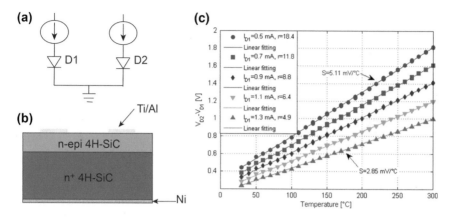

Fig. 2.19 **a** Temperature sensors using the SiC Schottky diode with **b** the schematic sketch of the device; **c** the linear sensing properties of 4H-SiC Schottky diode. Reprinted with permission from Ref. [79]

The thermoelectrical effect of Schottky 4H-SiC diodes has also employed for temperature sensing applications with a temperature of up to 300 °C [79]. Figures 2.19a, b show the schematic sketch of the Schottky sensor. For the working principle, two known currents I_{D1} and I_{D2} are applied while the voltage difference between two top electrodes is calculated as follows:

$$V_{D2} - V_{D1} = \frac{2k}{q} \eta \ln\left(\frac{I_{D2}}{I_{D1}}\right) + R_s(I_{D2} - I_{D1}) \tag{2.29}$$

where R_s is the parasitic resistance. The voltage difference is proportional to the temperature change as shown in Fig. 2.19c. From Eq. (2.29), the sensitivity of the device can be controlled by increasing the current ratio $r = I_{D2}/I_{D1}$. The maximum sensitivity was reported to be 5.11 mV/K at a current ratio of 18.4. The sensor showed excellent performance with good linearity and long-term stability.

PIN 4H-SiC diodes have been also used to develop highly sensitive temperature sensor [73] (Fig. 2.20a) with a maximum sensitivity of 2.66 mV/L at 10 μA applied current (Figs. 2.20b, c). Equation (2.26) was proven to govern the operation of the diode with an ideality factor of 2.05–2.45. The ideality factor increased with increasing the applied current.

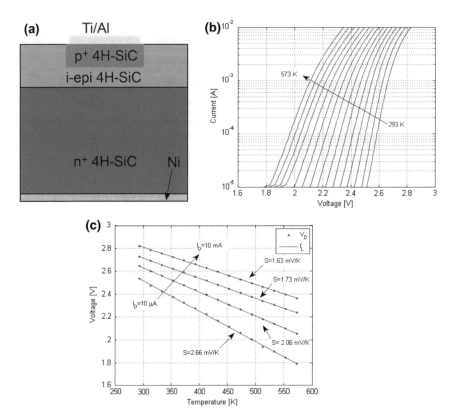

Fig. 2.20 PIN 4H-SiC temperature sensor. **a** Sensor structure. **b** Current–voltage characteristic of the sensor under temperature variation (293–573 K). **c** Sensitivity of the sensor. Reprinted with permission from Ref. [73]

2.7 Other Thermoelectrical Effects at High Temperatures

2.7.1 Thermoelectric Effect

Recently, a few Peltier modules of thermoelectric materials have been proposed to cool the power electronic devices. The performance of a thermoelectric material is evaluated by the figure of merit Z as follows [32]:

$$Z = \frac{S^2}{\rho k} \qquad (2.30)$$

where ρ is the resistivity of the thermoelectric material. To increase the cooling efficiency, the thermoelectric material should have a high Seebeck coefficient, a small resistivity and a low thermal conductivity [31–33].

Table 2.3 Seebeck coefficient (S) and thermal conductivity (k) of SiC in comparison to others [35]

Materials	Seebeck coefficient (S, μV/K)	Thermal conductivity (W/mK)
Bi_2Te_3	200	1.5
Si	200–1500	110
SiC	10–100	300
GaN	300–400	200
ZnO	450–550	100

Similar to other semiconductors, the temperature gradient in SiC can result in a potential difference or the Seebeck effect exists in SiC. SiC has been expected to be the next generation of thermoelectric material for harsh environments due to its excellent mechanical and electrical properties, as well as high melting point. Table 2.3 summarises the Seebeck coefficient S and thermal conductivity k of SiC comparison to other thermoelectric materials such as Si and Bi_2Te_3. For example, SiC has the smallest Seebeck coefficient value among these materials, while its thermal conductivity is highest. Therefore, the efficiency of using SiC as a thermoelectric material is relatively low in comparison with others, resulting in less successful demonstration of SiC for thermoelectric applications [35].

Fukuda et al. [32] investigated the thermoelectric properties of 4H-SiC and dense sintered SiC for self-cooling devices with a range of doping levels from 10^{16}–10^{19} cm^{-3}. The sintered SiC was fabricated using the compression of sintering fine-particle powders. Figure 2.21 shows the characterisation results for the Seebeck coefficients of different polytypes of SiC, indicating that 4H-SiC has a less Seebeck coefficient in comparison to others. For example, the Seebeck coefficient of 4H-SiC (e.g. less than 100 μV/K) is much lower than 6H-SiC (e.g. above 400 μV/K).

In addition, it has been reported that n-type SiC has a negative Seebeck coefficient, while p-type SiC has positive one (Fig. 2.22). Moreover, the Seebeck coefficient of 4H and 6H-SiC nanosheets slightly increases with increasing temperature [34], which is opposite to the report by Fukuda et al. [32]. In addition, the increase of the doping levels will lead to the decrease of the absolute Seebeck coefficient value. In these controversial phenomena, the impact of carrier concentration and temperature on the Seebeck coefficient cannot be explained by the typical semiconductor theory which attributes to the generation of the thermoelectromotive force under an application of temperature gradient. The phonon-drag effect should be taken into account, which is attributed to the predominant drags of carrier electrons by the phonons scattered by the stacking faults, boundaries and dislocations. Figure 2.22 shows the Seebeck coefficient of different types of SiC (e.g. 3C-SiC, 4H-SiC, 6H-SiC) which is from 400 to above 1000 for a temperature range up to 500 °C and different doping levels.

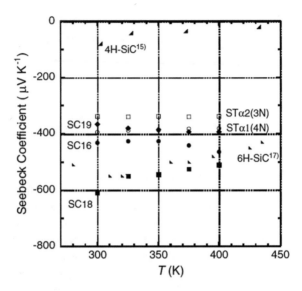

Fig. 2.21 Temperature dependence of Seebeck coefficient for sintered and single-crystalline SiC. Reprinted with permission from Ref. [32]

2.7.2 Thermocapacitive Effect

The depletion layer capacitance of a Schottky junction for the ideal cases varies with the voltage, which is given as follows [80, 81]:

$$C^{-2} = \frac{2(V_0 + V)}{q\varepsilon_s A^2 (N_d - N_a)} \tag{2.31}$$

where V_0 is the diffusion potential without applied bias voltage V; $N_d - N_a$ is the net ionised state density; q is the electric charge; A and ε_s are the area and permittivity of 4H-SiC. From Eq. (2.31), we have:

$$\frac{dC^{-2}}{dV} = \frac{2}{q\varepsilon_s A^2 (N_d - N_a)} \tag{2.32}$$

Recently, a few studies have investigated the temperature dependence of the thermocapacitance effect of metal/SiC Schottky junctions. For example, Fig. 2.23a shows the $1/C^2$-V characteristic of Au/4H-SiC structure in the temperature of 50–300 K [81]. The linear characteristic indicates the uniform dopant concentration of Au/4H-SiC. At the same bias voltage V, $1/C^2$ decreases with increasing temperature, corresponding to the increase of the capacitance C. It was explained that the net ionised

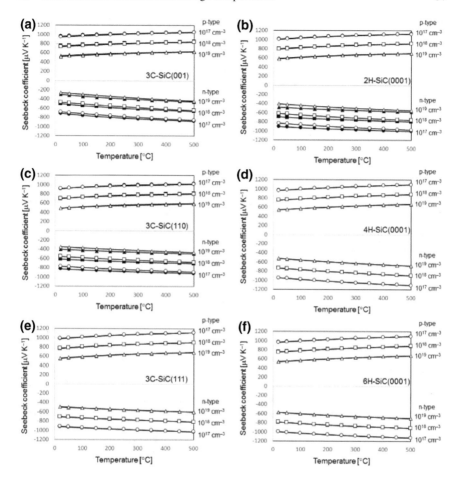

Fig. 2.22 Temperature dependence of Seebeck coefficient for different SiC polytypes and doping levels. Reprinted with permission from Ref. [34]

state density increased with increasing temperature from 6.58×10^{16} cm^{-3} at 50 K to 1.01×10^{17} cm^{-3} at 300 K. Figure 2.23b shows the increase of capacitance with increasing elevated temperature up to 160 °C, which is in solid agreement with the previous reported literature [80].

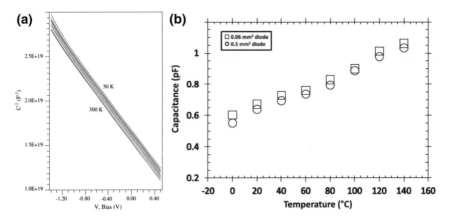

Fig. 2.23 a $1/C^2$-V characteristics of 4H-SiC junction for the temperature range of 50–300 K (Au/4H-SiC/Ni structure) [81]; **b** capacitance dependence on temperature on 4H-SiC UV Schottky photodiodes [80]. Reprinted with permission from Refs. [80, 81]

References

1. B. Verma, S. Sharma, Effect of thermal strains on the temperature coefficient of resistance. Thin Solid Films **5**, R44–R46 (1970)
2. P. Hall, The effect of expansion mismatch on temperature coefficient of resistance of thin films. Appl. Phys. Lett. **12**, 212–212 (1968)
3. T. Dinh, H.-P. Phan, A. Qamar, P. Woodfield, N.-T. Nguyen, D.V. Dao, Thermoresistive effect for advanced thermal sensors: fundamentals, design considerations, and applications. J. Microelectromech. Syst. (2017)
4. J.T. Kuo, L. Yu, E. Meng, Micromachined thermal flow sensors—a review. Micromachines **3**, 550–573 (2012)
5. F. Warkusz, The size effect and the temperature coefficient of resistance in thin films. J. Phys. D Appl. Phys. **11**, 689 (1978)
6. V.T. Dau, D.V. Dao, T. Shiozawa, H. Kumagai, S. Sugiyama, Development of a dual-axis thermal convective gas gyroscope. J. Micromech. Microeng. **16**, 1301 (2006)
7. T. Dinh, H.-P. Phan, T. Kozeki, A. Qamar, T. Namazu, N.-T. Nguyen et al., Thermoresistive properties of p-type 3C–SiC nanoscale thin films for high-temperature MEMS thermal-based sensors. RSC Adv **5**, 106083–106086 (2015)
8. S.O. Kasap, *Principles of Electronic Materials and Devices* (McGraw-Hill, New York, 2006)
9. S.M. Sze, K.K. Ng, *Physics of Semiconductor Devices* (Wiley, New York, 2006
10. T. Dinh, H.-P. Phan, A. Qamar, P. Woodfield, N.-T. Nguyen, D.V. Dao, Thermoresistive effect for advanced thermal sensors: fundamentals, design considerations, and applications. J. Microelectromech. Syst. **26**, 966–986 (2017)
11. T. Dinh, D.V. Dao, H.-P. Phan, L. Wang, A. Qamar, N.-T. Nguyen et al., Charge transport and activation energy of amorphous silicon carbide thin film on quartz at elevated temperature. Appl. Phys. Express **8**, 061303 (2015)
12. K. Sasaki, E. Sakuma, S. Misawa, S. Yoshida, S. Gonda, High-temperature electrical properties of 3C-SiC epitaxial layers grown by chemical vapor deposition. Appl. Phys. Lett. **45**, 72–73 (1984)
13. T. Nagai, K. Yamamoto, I. Kobayashi, Rapid response SiC thin-film thermistor. Rev. Sci. Instrum. **55**, 1163–1165 (1984)
14. T. Nagai, M. Itoh, SiC thin-film thermistors. IEEE Trans. Ind. Appl. **26**, 1139–1143 (1990)

15. E.A. de Vasconcelos, S. Khan, W. Zhang, H. Uchida, T. Katsube, Highly sensitive thermistors based on high-purity polycrystalline cubic silicon carbide. Sens. Actuators, A **83**, 167–171 (2000)
16. T. Dinh, H.-P. Phan, T. Kozeki, A. Qamar, T. Fujii, T. Namazu et al., High thermosensitivity of silicon nanowires induced by amorphization. Mater. Lett. **177**, 80–84 (2016)
17. J.Y. Seto, The electrical properties of polycrystalline silicon films. J. Appl. Phys. **46**, 5247–5254 (1975)
18. N.-C. Lu, L. Gerzberg, C.-Y. Lu, J.D. Meindl, A conduction model for semiconductor-grain-boundary-semiconductor barriers in polycrystalline-silicon films. IEEE Trans. Electron Devices **30**, 137–149 (1983)
19. D.M. Kim, A. Khondker, S. Ahmed, R.R. Shah, Theory of conduction in polysilicon: drift-diffusion approach in crystalline-amorphous-crystalline semiconductor system—Part I: Small signal theory. IEEE Trans. Electron Devices **31**, 480–493 (1984)
20. A. Singh, Grain-size dependence of temperature coefficient of resistance of polycrystalline metal films. Proc. IEEE **61**, 1653–1654 (1973)
21. J.T. Irvine, A. Huanosta, R. Valenzuela, A.R. West, Electrical properties of polycrystalline nickel zinc ferrites. J. Am. Ceram. Soc. **73**, 729–732 (1990)
22. A. Singh, Film thickness and grain size diameter dependence on temperature coefficient of resistance of thin metal films. J. Appl. Phys. **45**, 1908–1909 (1974)
23. S. Baranovski, *Charge Transport in Disordered Solids with Applications in Electronics*, vol. 17 (Wiley, New York, 2006)
24. N.F. Mott, E.A. Davis, *Electronic Processes in Non-Crystalline Materials* (OUP Oxford, 2012)
25. R. Street, *Hydrogenated Amorphous Silicon* (Cambridge University, Cambridge, 1991)
26. P. Fenz, H. Muller, H. Overhof, P. Thomas, Activated transport in amorphous semiconductors. II. Interpretation of experimental data. J. Phys. C: Solid State Phys. **18**, 3191 (1985)
27. D. Peters, R. Schörner, K.-H. Hölzlein, P. Friedrichs, Planar aluminum-implanted 1400 V 4H silicon carbide pn diodes with low on resistance. Appl. Phys. Lett. **71**, 2996–2997 (1997)
28. Y.S. Ju, Analysis of thermocapacitive effects in electric double layers under a size modified mean field theory. Appl. Phys. Lett. **111**, 173901 (2017)
29. C.Y. Kwok, K.M. Lin, R.S. Huang, A silicon thermocapacitive flow sensor with frequency modulated output. Sens. Actuators, A **57**, 35–39 (1996)
30. N. Abu-Ageel, M. Aslam, R. Ager, L. Rimai, The Seebeck coefficient of monocrystalline-SiC and polycrystalline-SiC measured at 300–533 K. Semicond. Sci. Technol. **15**, 32 (2000)
31. C.-H. Pai, Thermoelectric properties of p-type silicon carbide, in *XVII International Conference on Thermoelectrics, 1998. Proceedings ICT 98* (1998), pp. 582–586
32. S. Fukuda, T. Kato, Y. Okamoto, H. Nakatsugawa, H. Kitagawa, S. Yamaguchi, Thermoelectric properties of single-crystalline SiC and dense sintered SiC for self-cooling devices. Jpn. J. Appl. Phys. **50**, 031301 (2011)
33. P. Wang, Recent advance in thermoelectric devices for electronics cooling, in *Encyclopedia of Thermal Packaging: Thermal Packaging Tools* (World Scientific, 2015), pp. 145–168
34. K. Nakamura, First-principles simulation on Seebeck coefficient in silicon and silicon carbide nanosheets. Jpn. J. Appl. Phys. **55**, 06GJ07 (2016)
35. Y. Furubayashi, T. Tanehira, A. Yamamoto, K. Yonemori, S. Miyoshi, S.-I. Kuroki, Peltier effect of silicon for cooling 4H-SiC-based power devices. ECS Trans. **80**, 77–85 (2017)
36. T.-K. Nguyen, H.-P. Phan, T. Dinh, T. Toriyama, K. Nakamura, A.R.M. Foisal et al., Isotropic piezoresistance of p-type 4H-SiC in (0001) plane. Appl. Phys. Lett. **113**, 012104 (2018)
37. A.R.M. Foisal, T. Dinh, P. Tanner, H.-P. Phan, T.-K. Nguyen, E.W. Streed et al., Photoresponse of a highly-rectifying 3C-SiC/Si heterostructure under UV and visible illuminations. IEEE Electron Device Lett. (2018)
38. T. Dinh, H.-P. Phan, T.-K. Nguyen, V. Balakrishnan, H.-H. Cheng, L. Hold et al., Unintentionally doped epitaxial 3C-SiC (111) nanothin film as material for highly sensitive thermal sensors at high temperatures. IEEE Electron Device Lett. **39**, 580–583 (2018)
39. J.S. Shor, D. Goldstein, A.D. Kurtz, Characterization of n-type beta-SiC as a piezoresistor. IEEE Trans. Electron Devices **40**, 1093–1099 (1993)

40. J.S. Shor, L. Bemis, A.D. Kurtz, Characterization of monolithic n-type 6H-SiC piezoresistive sensing elements. IEEE Trans. Electron Devices **41**, 661–665 (1994)
41. R.S. Okojie, A.A. Ned, A.D. Kurtz, W.N. Carr, Characterization of highly doped n-and p-type 6H-SiC piezoresistors. IEEE Trans. Electron Devices **45**, 785–790 (1998)
42. T. Abtew, M. Zhang, D. Drabold, Ab initio estimate of temperature dependence of electrical conductivity in a model amorphous material: hydrogenated amorphous silicon. Phys. Rev. B **76**, 045212 (2007)
43. E.A. de Vasconcelos, W.Y. Zhang, H. Uchida, T. Katsube, Potential of high-purity polycrystalline silicon carbide for thermistor applications. Jpn. J. Appl. Phys. **37**, 5078 (1998)
44. K. Wasa, T. Tohda, Y. Kasahara, S. Hayakawa, Highly-reliable temperature sensor using rf-sputtered SiC thin film. Rev. Sci. Instrum. **50**, 1084–1088 (1979)
45. T. Dinh, H.-P. Phan, N. Kashaninejad, T.-K. Nguyen, D.V. Dao, N.-T. Nguyen, An on-chip SiC MEMS device with integrated heating, sensing and microfluidic cooling systems. Adv. Mater. Interfaces **1**, 1 (2018)
46. H.-P. Phan, D.V. Dao, P. Tanner, L. Wang, N.-T. Nguyen, Y. Zhu et al., Fundamental piezoresistive coefficients of p-type single crystalline 3C-SiC. Appl. Phys. Lett. **104**, 111905 (2014)
47. S.S. Li, *The dopant density and temperature dependence of electron mobility and resistivity in n-type silicon*. US Dept. of Commerce, National Bureau of Standards; for sale by the Supt. of Docs., US Govt. Print. Off. (1977)
48. M. Roschke, F. Schwierz, Electron mobility models for 4H, 6H, and 3C SiC [MESFETs]. IEEE Trans. Electron Devices **48**, 1442–1447 (2001)
49. R. Humphreys, D. Bimberg, W. Choyke, Wavelength modulated absorption in SiC. Solid State Commun. **39**, 163–167 (1981)
50. H. Mukaida, H. Okumura, J. Lee, H. Daimon, E. Sakuma, S. Misawa et al., Raman scattering of SiC: estimation of the internal stress in 3C-SiC on Si. J. Appl. Phys. **62**, 254–257 (1987)
51. M. Wieligor, Y. Wang, T. Zerda, Raman spectra of silicon carbide small particles and nanowires. J. Phys.: Condens. Matter **17**, 2387 (2005)
52. A. Qamar, H.-P. Phan, J. Han, P. Tanner, T. Dinh, L. Wang et al., The effect of device geometry and crystal orientation on the stress-dependent offset voltage of 3C–SiC (100) four terminal devices. J. Mater. Chem. C **3**, 8804–8809 (2015)
53. H.-P. Phan, H.-H. Cheng, T. Dinh, B. Wood, T.-K. Nguyen, F. Mu et al., Single-crystalline 3C-SiC anodically bonded onto glass: an excellent platform for high-temperature electronics and bioapplications. ACS Appl. Mater. Interfaces. **9**, 27365–27371 (2017)
54. M.S. Raman, T. Kifle, E. Bhattacharya, K. Bhat, Physical model for the resistivity and temperature coefficient of resistivity in heavily doped polysilicon. IEEE Trans. Electron Devices **53**, 1885–1892 (2006)
55. M. Yamanaka, H. Daimon, E. Sakuma, S. Misawa, S. Yoshida, Temperature dependence of electrical properties of n-and p-type 3C-SiC. J. Appl. Phys. **61**, 599–603 (1987)
56. M. Yamanaka, K. Ikoma, Temperature dependence of electrical properties of 3C-SiC (1 1 1) heteroepitaxial films. Physica B **185**, 308–312 (1993)
57. X. Song, J. Michaud, F. Cayrel, M. Zielinski, M. Portail, T. Chassagne et al., Evidence of electrical activity of extended defects in 3C–SiC grown on Si. Appl. Phys. Lett. **96**, 142104 (2010)
58. C. Yan, J. Wang, P.S. Lee, Stretchable graphene thermistor with tunable thermal index. ACS Nano **9**, 2130–2137 (2015)
59. D. Kong, L.T. Le, Y. Li, J.L. Zunino, W. Lee, Temperature-dependent electrical properties of graphene inkjet-printed on flexible materials. Langmuir **28**, 13467–13472 (2012)
60. Q. Gao, H. Meguro, S. Okamoto, M. Kimura, Flexible tactile sensor using the reversible deformation of poly (3-hexylthiophene) nanofiber assemblies. Langmuir **28**, 17593–17596 (2012)
61. X. She, A.Q. Huang, Ó. Lucía, B. Ozpineci, Review of silicon carbide power devices and their applications. IEEE Trans. Industr. Electron. **64**, 8193–8205 (2017)
62. G.L. Harris, *Properties of Silicon Carbide* (IET, 1995)

63. H.-P. Phan, T. Dinh, T. Kozeki, T.-K. Nguyen, A. Qamar, T. Namazu et al., The piezoresistive effect in top-down fabricated p-type 3C-SiC nanowires. IEEE Electron Device Lett. **37**, 1029–1032 (2016)

64. L. Wang, S. Dimitrijev, J. Han, A. Iacopi, L. Hold, P. Tanner et al., Growth of 3C–SiC on 150-mm Si (100) substrates by alternating supply epitaxy at 1000 °C. Thin Solid Films **519**, 6443–6446 (2011)

65. L. Wang, A. Iacopi, S. Dimitrijev, G. Walker, A. Fernandes, L. Hold et al., Misorientation dependent epilayer tilting and stress distribution in heteroepitaxially grown silicon carbide on silicon (111) substrate. Thin Solid Films **564**, 39–44 (2014)

66. V. Afanas'ev, M. Bassler, G. Pensl, M. Schulz, E. Stein von Kamienski, Band offsets and electronic structure of SiC/SiO_2 interfaces. J. Appl. Phys. **79**, 3108–3114 (1996)

67. P. Tanner, S. Dimitrijev, H.B. Harrison, Current mechanisms in n-SiC/p-Si heterojunctions, in *Conference on Optoelectronic and Microelectronic Materials and Devices, 2008. COMMAD 2008* (2008), pp. 41–43

68. A. Qamar, P. Tanner, D.V. Dao, H.-P. Phan, T. Dinh, Electrical properties of p-type 3C-SiC/Si heterojunction diode under mechanical stress. IEEE Electron Device Lett. **35**, 1293–1295 (2014)

69. S.Z. Karazhanov, I. Atabaev, T. Saliev, É. Kanaki, E. Dzhaksimov, Excess tunneling currents in p-Si-n-3C-SiC heterostructures. Semiconductors **35**, 77–79 (2001)

70. P. Yih, J. Li, A. Steckl, SiC/Si heterojunction diodes fabricated by self-selective and by blanket rapid thermal chemical vapor deposition. IEEE Trans. Electron Devices **41**, 281–287 (1994)

71. L. Marsal, J. Pallares, X. Correig, A. Orpella, D. Bardés, R. Alcubilla, Analysis of conduction mechanisms in annealed n-Si $1 - x$ C x: H/p-crystalline Si heterojunction diodes for different doping concentrations. J. Appl. Phys. **85**, 1216–1221 (1999)

72. S.B. Hou, P.E. Hellström, C.M. Zetterling, M. Östling, 4H-SiC PIN diode as high temperature multifunction sensor, in *Materials Science Forum* (2017, pp. 630–633)

73. S. Rao, G. Pangallo, F.G. Della Corte, 4H-SiC pin diode as highly linear temperature sensor. IEEE Trans. Electron Devices **63**, 414–418 (2016)

74. G. Brezeanu, M. Badila, F. Draghici, R. Pascu, G. Pristavu, F. Craciunoiu, et al., High temperature sensors based on silicon carbide (SiC) devices, in *2015 International Semiconductor Conference (CAS)* (2015), pp. 3–10.

75. S. Rao, G. Pangallo, F.G. Della Corte, Highly linear temperature sensor based on 4H-silicon carbide pin diodes. IEEE Electron Device Lett. **36**, 1205–1208 (2015)

76. V. Cimalla, J. Pezoldt, O. Ambacher, Group III nitride and SiC based MEMS and NEMS: materials properties, technology and applications. J. Phys. D Appl. Phys. **40**, 6386 (2007)

77. M. Mehregany, C.A. Zorman, SiC MEMS: opportunities and challenges for applications in harsh environments. Thin Solid Films **355**, 518–524 (1999)

78. N. Zhang, C.-M. Lin, D.G. Senesky, A.P. Pisano, Temperature sensor based on 4H-silicon carbide pn diode operational from 20 °C to 600 °C. Appl. Phys. Lett. **104**, 073504 (2014)

79. S. Rao, G. Pangallo, F. Pezzimenti, F.G. Della Corte, High-performance temperature sensor based on 4H-SiC Schottky diodes. IEEE Electron Device Lett. **36**, 720–722 (2015)

80. S. Zhao, G. Lioliou, A. Barnett, Temperature dependence of commercial 4H-SiC UV Schottky photodiodes for X-ray detection and spectroscopy. Nucl. Instrum. Methods Phys. Res., Sect. A **859**, 76–82 (2017)

81. M. Gülnahar, Temperature dependence of current-and capacitance–voltage characteristics of an Au/4H-SiC Schottky diode. Superlattices Microstruct. **76**, 394–412 (2014)

Chapter 3
Desirable Features for High-Temperature SiC Sensors

Abstract There are several parameters representing the detection capability and efficiency of SiC sensors towards practical applications such as temperature detectors, flow sensors, convective-based accelerometers and gyroscopes. This chapter presents a number of desirable parameters for SiC thermal sensors at high temperatures. These important features include, but are not limited to, the sensitivity, response time and linearity.

Keywords Sensitivity · Response time · Power consumption · Stability

3.1 Sensitivity

Since the detectability and high efficiency are the main features of electronic sensing devices, the sensitivity is considered as one of the most important parameters for the development of SiC sensors [1]. As discussed in Chap. 2, the sensitivity of SiC thermal sensors is evaluated using the temperature coefficient of resistance (TCR), defined by the electrical resistance change to the temperature difference. The TCR is calculated as: $\mathrm{TCR} = \Delta R/R_o \times 1/\Delta T$, where $\Delta R/R_o = (R - R_o)/R_o$ and $\Delta T = T - T_o$ are the resistance change and temperature change [2], respectively.

A large TCR value is preferred for highly sensitive temperature sensors; thus, ceramic or composite materials are typically utilised owing to their high thermoresistive sensitivity with a TCR of up to 10^{12} ppm/K [3]. However, due to the high resistivity, these materials might not be a material choice for Joule heating-based thermal MEMS sensors, such as thermal flow sensors and convective inertial sensors (e.g. accelerometers and gyroscopes) [4–7]. This is due to the fact that Joule heating-based sensors employ low resistivity materials to reduce the voltage or current supply [8]. Therefore, conventional materials such as metals and highly doped semiconductors have been typically used to fabricate Joule heating-based MEMS sensors [9–11]. However, these materials have low thermoresistive sensitivity with a TCR value of less than 5000 ppm/K [12, 13].

© The Author(s) 2018
T. Dinh et al., *Thermoelectrical Effect in SiC for High-Temperature MEMS Sensors*, SpringerBriefs in Applied Sciences and Technology, https://doi.org/10.1007/978-981-13-2571-7_3

43

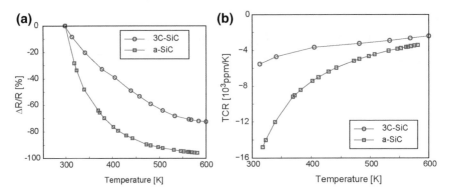

Fig. 3.1 SiC thermoresistive sensitivity [2, 18]. **a** Relative resistance change of SiC nanofilms. **b** Temperature coefficient of resistance (TCR)

SiC has been demonstrated as a potential material for thermal sensing applications at high temperatures [14–17]. Amorphous SiC films offer high sensitivity up to −16,000 ppm/K, while the highly doped single-crystalline configuration has a lower sensitivity [1, 2, 18–20]. Figure 3.1a shows the decrease in the electrical resistance with increasing temperature. This corresponds to an increase in the absolute relative resistance change which is approximately 70 and 95% for single-crystalline 3C-SiC and amorphous SiC, respectively. Consequently, high negative TCR values of −2000 to −6000 ppm/K and −4000 to −16,000 ppm/K [18] (Fig. 3.1b) were reported for 3C-SiC and a-SiC [4, 5], respectively. This high sensitivity demonstrates strong feasibility of using SiC nanofilms for thermal sensors. However, the high sensitivity of SiC at a wider and higher temperature range (500–800 K) is desirable for thermal sensors operating in high-temperature environments.

Apart from the TCR and resistivity considerations, the simplicity in design and ease of microfabrication processing are also important factors for Joule heating-based sensors. In addition, the packaging of such sensors will affect the selection of materials as direct exposure to corrosive environments leads to the degradation of sensor performance over time [15, 17, 21–23].

For thermal flow sensors employing high TCR materials, the measurement can be simplified by directly measuring the resistance change of sensing elements [24, 25]. Figure 3.2a shows the direct measurement of the resistance change used for the materials with very high TCR. Since conventional thermal sensors employ low TCR materials, the resistance change (ΔR) of the sensors with applying input parameters is typically converted to a voltage change (ΔV) using a Wheatstone bridge and then amplified with a signal amplifier. Figure 3.2b shows a schematic sketch of an electrical circuit including a Wheatstone bridge used for calorimetric flow sensors [26, 27].

Alternatively, for temperature detectors employing ceramic and semiconductors, the thermal index B (K), are deployed as the sensitivity. The thermal index B can be

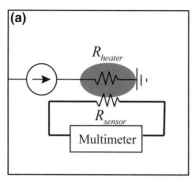
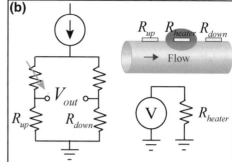

Direct measurement
of sensor resistance

In direct measurement using a
Wheatstone bridge

Fig. 3.2 a Direct measurement of resistance change. **b** Measurement of resistance change using a Wheatstone bridge

Table 3.1 Thermal index or thermistor coefficient of SiC [15, 17, 21–23, 28]

Material	Technique	Temperature range (K)	Thermal index B (K)
a-SiC	LPCVD	300–580	1750–2400
SiC/diamond/Si	MPCVD	300–570	550–4500
3C–SiC	Rf-sputtering	200–720	1600–3400
3C–SiC	Rf-sputtering	275–770	2000–4000
3C–SiC	CVD	300–670	5000–7000

defined based on the temperature coefficient of resistance (TCR) as $B = \text{TCR} \times T^2$. Table 3.1 shows the thermal index B of various SiC materials reported in the literature.

In addition to the application of a single layer of SiC for thermal sensing, multi-layers of SiC have also been employed as temperature sensors. The thermoelectronic sensitivity of SiC p–n junctions is defined as the voltage change with temperature change (e.g. dV/dT) under a constant applied current [29]. Currently, there have been a few reports on the utilisation of p–n 4H–SiC junctions for temperature sensing applications up to 600 °C, with a high sensitivity of up to 5 mV/K [30–35]. However, very limited work has demonstrated the high performance of the sensors for high-temperature applications.

3.2 Linearity

A linear characteristic is an important feature of SiC thermal sensors as it represents the measurement accuracy and design simplicity of circuitry. The linear characteristic typically refers to the linear relationship between the resistance and the temperature as follows [26]:

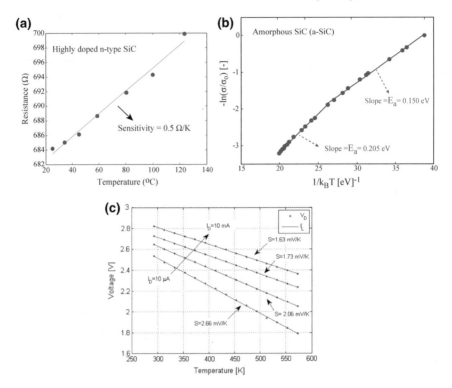

Fig. 3.3 Linear characteristic of thermoresistive effect in SiC films. **a** Highly doped n-type 3C–SiC. **b** Linear Arrhenius plot of amorphous SiC thermoresistance [18]. **c** Linear characteristic of p–n 4H–SiC junction diodes [34]. Reprinted with permission from Refs. [18, 34, 36]

$$R = R_o[1 + \alpha(T - T_o)] \qquad (3.1)$$

where R and R_o are the resistance at temperature T and T_o, respectively. α denotes the temperature coefficient of resistance TCR. Figure 3.3a shows the linear change of the electrical resistance with increasing temperature. From the linear fit in Fig. 3.3a, a temperature sensitivity of 0.5 Ω/K is measured for the highly doped SiC films. To employ SiC films as temperature sensors, the temperature increases 43 K for every 1% of resistance change. Therefore, the temperature in highly doped n-type SiC films can be described as $T = 25\,°C + 43 \times \Delta R/R$ (%). However, the electrical resistance of semiconductors is exponentially dependent on temperature as [20, 26, 37, 38]:

$$R = A \, \exp\left(\frac{E_a}{kT}\right) \qquad (3.2)$$

where E_a is the activation energy and k is the Boltzmann constant. The linear relationship between $\ln(R/R_o)$ and $1/T$ can be derived from Eq. (3.2) as follows:

$$\ln(R/R_o) = A - B\left(\frac{1}{T}\right) \tag{3.3}$$

where A and B are constants. The thermoresistive sensitivity of the material B is determined from the slope of the plot $\ln(R/R_o)$ versus $1/T$. It is important to note that the activation energy E_a is determined by $E_a = B \times k$. Figure 3.3b shows an example of linear fitting of the thermoresistive effect in amorphous SiC films. From the slope, the activation energy is determined to be 150 and 205 meV for the temperature range of 300–450 and 450–580 K [18], respectively.

For the application of the thermoelectrical effect in SiC junctions, a number of studies have demonstrated the linearity of the sensor response for a temperature of up to 300 °C. Figure 3.3c shows the linear sensitivity of the 4H–SiC p–n junction for a temperature up to 573 K. At a constant applied current of 10 μA, the highest sensitivity of 2.66 mV/K was achieved [34]. Increasing the applied current leads to a decrease of the sensitivity.

3.3 Thermal Time Response

A fast response of SiC thermal sensors is desired for high-performance MEMS sensing devices as the thermal time response presents the instantaneous or rapid response to external inputs including temperature, stress/strain and acceleration. Experimentally, the thermal response time is calculated as the amount of time for the response signal to reach 63.2 or 90% the steady-state signal [27, 39–43] (Fig. 3.4a). The thermal response time can also be defined from the exponential fit for the temperature response of thermal sensors as follows [40, 44–47].

Figure 3.4a shows thermal time response of the 63.2 or 90% time to reach a steady-state signal. Figure 3.4b shows thermal time response determined by fitting.

$$T = A - B \exp\left(\frac{-t}{\tau}\right) \tag{3.4}$$

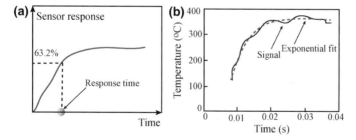

Fig. 3.4 Response time of SiC thermal sensors. **a** 63.2% response time of SiC thermal sensors. **b** Determination of the thermal response time based on the fitting lineshape [26, 39, 40]

where A and B are constants; t is the time parameter; and τ is the thermal response time. Figure 3.4b shows the determination of 2.5 ms for the thermal time response of a SiC thermal flow sensor [40].

Theoretically, the thermal time response is limited by the thermal time constant $\tau = R_{th} \times C_{th}$, where R_{th} and C_{th} are the thermal resistance and thermal capacitance. The thermal resistance is calculated as:

$$R_{th} = \frac{L}{A \times k} \tag{3.5}$$

where L, A and k are the length, cross-sectional area and thermal conductivity. To achieve a faster time response, a lower thermal resistance should be designed. For example, the utilisation of a sensor material with a high thermal conductivity k will result in a fast time response. In addition, the thermal capacitance is expressed in the following form:

$$C_{th} = \rho \times V \times C_H \tag{3.6}$$

where ρ and C_H are the density and heat capacity of the material, respectively; $V = L \times A$ is the volume of the sensor. Therefore, the time constant is determined as:

$$\tau = R_{th} \times C_{th} = \frac{L}{A \times k} \times l \times A \times V \times C_H = L^2 \times \rho \times C_H / k \tag{3.7}$$

In general, materials with low density, low heat capacity and high thermal conductivity should be employed to achieve a fast response time. As an example, platinum, a common material for thermal sensors, has a low heat capacity of 125 J/(kgK), while its density is relatively high (e.g. 21.45 g/cm^3). Silicon and silicon carbide have a lower density of 2.33 and 3.2 g/cm^3, respectively, but a high heat capacity of approximately 700 J/(kgK). Therefore, scaling down the size of thermal sensors is an alternative strategy to achieve fast thermal time response. Nanoscale systems are of high interest for measuring the response of systems to rapid environment temperature change. Using Eq. (3.7), the thermal time constant is calculated to be less than 20 ms for SiC bridges with a length of 1000 μm [26, 40].

3.4 Low Power Consumption

Low power consumption is particularly relevant for electronic systems including thermal sensors. It is well known that the sensitivity of thermal sensors increases with increasing applied power. Therefore, thermal sensors employing high TCR materials will lower the power consumption while maintaining high output signals. Miniaturisation of thermal sensors using micro-/nanomachining technologies can lead to a low power consumption of such sensors. A small heater will consume less power to increase the temperature to a certain steady-state temperature and improve the response time of thermal sensors. At a constant power consumption, the sensitivity of thermal sensors will increase when the size of sensor structures is scaled down. Therefore, one-dimensional (1D) and two-dimensional (2D) materials will be of interest for thermal sensors with low power consumption and fast thermal response [25, 43, 48].

For temperature sensors, in which heating components are not required, the power consumption is typically low [43, 48, 49]. The response time and uniformity of temperature distribution on the sensors can be also improved by scaling down the length, width and thickness of such sensors [50–55]. However, due to the decrease of the thin film quality and change of material morphologies, the thermosensitivity decreases with increasing film thickness [56, 57].

For Joule heating-based sensors constructed on a thermally conductive substrate (e.g. SiC film grown on Si substrate), power loss is relatively high due to the heat conduction to the substrate. It is required for bonding of SiC films on an insulation substrate, such as glass or SOI wafer (silicon on insulator) substrates, and subsequently removing the conducting Si substrate. Another solution is to design a longer and wider heater to reduce the power loss. However, a longer heating element requires a higher applied voltage as it has a higher resistance. In addition, optimisation of distance between the heating and sensing elements will boost the sensitivity of such sensing devices.

Conventionally, fabrication of heating elements on an isolation substrate or using released structures will eliminate the conduction loss and reduce the power consumption overall. To fabricate the suspended structures, wet etching processes can be employed to release the functional heating and sensing elements from the substrate. These strategies can lead to the development of sensing devices with a low power consumption of less than 1 mW with large measurement range, high resolution and fast thermal time response [43, 48, 49]. It is noteworthy that the decrease of supply power results in a lower sensitivity of thermal sensors. Consequently, a large thermal–electrical effect is desired for highly sensitive, high-resolution thermal sensors working at low power consumption.

3.5 Stability and Other Desirable Features

It is well known that the performance of thermal sensing device will decline when working in harsh environments under long-term service. Therefore, the consideration of repeatability and long-term stability of thermal sensing devices is an important design task. The repeatability characteristics are typically determined from the repeated cycle tests to prove the stability of the sensor response to the input signal. For example, Fig. 3.5a shows the repeated response of a SiC thermal flow sensor under the application of a flow velocity of 2.1 m/s [25]. Moreover, the heat test was performed for SiC thermistor from −196 to 550 °C (each test 10 s), and the properties of the sensor were not changed with 10,000 cycles. In addition, SiC films

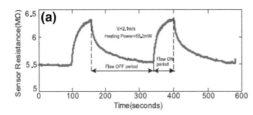 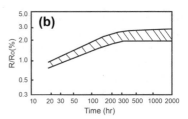

Fig. 3.5 **a** Response of a SiC thermal flow sensor under two repeated cycles of flow rates [25]. **b** Long-term stability of SiC sensors [14]. Reprinted with permission from Refs. [14, 25]

were demonstrated as a thermistor with good long-term stability when continuously exposed to high temperature (~400 °C) in air for 2000 h (Fig. 3.5b). The resistance change was measured to be 0.6% at 350 °C and 5% at 450 °C [15, 17, 22].

In addition, MEMS thermal sensors typically operate in outdoor conditions with uncontrolled humidity and chemical species, resulting in the instability of the sensor response [58, 59]. Highly porous materials adsorb humidity and frequently suffer the drift of the sensor response. Therefore, low porosity materials are desired for lowering the effect on humidity on thermoresistive sensors. SiC sheet can be used as humidity sensors, while SiC thick films will not be affected significantly by humidity.

References

1. T. Dinh, H.-P. Phan, A. Qamar, P. Woodfield, N.-T. Nguyen, D. V. Dao, Thermoresistive effect for advanced thermal sensors: fundamentals, design considerations, and applications. J. Microelectromech. Syst. (2017)
2. T. Dinh, H.-P. Phan, T. Kozeki, A. Qamar, T. Namazu, N.-T. Nguyen et al., Thermoresistive properties of p-type 3C–SiC nanoscale thin films for high-temperature MEMS thermal-based sensors. RSC Adv. **5**, 106083–106086 (2015)
3. K. Ohe, Y. Naito, A new resistor having an anomalously large positive temperature coefficient. Jpn. J. Appl. Phys. **10**, 99 (1971)
4. J. Bahari, J.D. Jones, A.M. Leung, Sensitivity improvement of micromachined convective accelerometers. J. Microelectromech. Syst. **21**, 646–655 (2012)
5. V.T. Dau, D.V. Dao, T. Shiozawa, S. Sugiyama, Simulation and fabrication of a convective gyroscope. IEEE Sens. J. **8**, 1530–1538 (2008)
6. V.T. Dau, T. Yamada, D.V. Dao, B.T. Tung, K. Hata, S. Sugiyama, Integrated CNTs thin film for MEMS mechanical sensors. Microelectron. J. **41**, 860–864 (2010)
7. V.T. Dau, B.T. Tung, T.X. Dinh, D.V. Dao, T. Yamada, K. Hata et al., A micromirror with CNTs hinge fabricated by the integration of CNTs film into a MEMS actuator. J. Micromech. Microeng. **23**, 075024 (2013)
8. A. Vatani, P.L. Woodfield, T. Dinh, H.-P. Phan, N.-T. Nguyen, D.V. Dao, Degraded boiling heat transfer from hotwire in ferrofluid due to particle deposition. Appl. Therm. Eng. (2018)
9. V.T. Dau, D.V. Dao, T. Shiozawa, H. Kumagai, S. Sugiyama, Development of a dual-axis thermal convective gas gyroscope. J. Micromech. Microeng. **16**, 1301 (2006)

10. V.T. Dau, D.V. Dao, S. Sugiyama, A 2-DOF convective micro accelerometer with a low thermal stress sensing element, in *Based on work presented at IEEE sensor 2006: the 5th IEEE conference on sensors*, Daegu, Korea, 22–25 Oct 2006. Smart Mater. Struct. **16**, 2308 (2007)
11. D.V. Dao, V.T. Dau, T. Shiozawa, S. Sugiyama, Development of a dual-axis convective gyroscope with low thermal-induced stress sensing element. J. Microelectromech. Syst. **16**, 950 (2007)
12. A.M. Leung, J. Jones, E. Czyzewska, J. Chen, M. Pascal, Micromachined accelerometer with no proof mass, in *International Electron Devices Meeting, 1997. IEDM '97. Technical Digest*, 1997, pp. 899–902
13. J.T. Kuo, L. Yu, E. Meng, Micromachined thermal flow sensors—a review. Micromachines **3**, 550–573 (2012)
14. K. Wasa, T. Tohda, Y. Kasahara, S. Hayakawa, Highly-reliable temperature sensor using rf-sputtered SiC thin film. Rev. Sci. Instrum. **50**, 1084–1088 (1979)
15. T. Nagai, K. Yamamoto, I. Kobayashi, Rapid response SiC thin-film thermistor. Rev. Sci. Instrum. **55**, 1163–1165 (1984)
16. K. Sasaki, E. Sakuma, S. Misawa, S. Yoshida, S. Gonda, High-temperature electrical properties of 3C–SiC epitaxial layers grown by chemical vapor deposition. Appl. Phys. Lett. **45**, 72–73 (1984)
17. T. Nagai, M. Itoh, SiC thin-film thermistors. IEEE Trans. Ind. Appl. **26**, 1139–1143 (1990)
18. T. Dinh, D.V. Dao, H.-P. Phan, L. Wang, A. Qamar, N.-T. Nguyen et al., Charge transport and activation energy of amorphous silicon carbide thin film on quartz at elevated temperature. Appl. Phys. Express **8**, 061303 (2015)
19. A.R.M. Foisal, H.-P. Phan, T. Kozeki, T. Dinh, K.N. Tuan, A. Qamar et al., 3C–SiC on glass: an ideal platform for temperature sensors under visible light illumination. RSC Adv. **6**, 87124–87127 (2016)
20. T. Dinh, H.-P. Phan, T.-K. Nguyen, V. Balakrishnan, H.-H. Cheng, L. Hold et al., Unintentionally doped epitaxial 3C–SiC (111) nanothin film as material for highly sensitive thermal sensors at high temperatures. IEEE Electron Device Lett. **39**, 580–583 (2018)
21. E.A. de Vasconcelos, W.Y. Zhang, H. Uchida, T. Katsube, Potential of high-purity polycrystalline silicon carbide for thermistor applications. Jpn. J. Appl. Phys. **37**, 5078 (1998)
22. E.A. de Vasconcelos, S. Khan, W. Zhang, H. Uchida, T. Katsube, Highly sensitive thermistors based on high-purity polycrystalline cubic silicon carbide. Sens. Actuators A Phys. **83**, 167–171 (2000)
23. N. Boltovets, V. Kholevchuk, R. Konakova, Y.Y. Kudryk, P. Lytvyn, V. Milenin et al., A silicon carbide thermistor. Semicond. Phys. Quantum Electron. Optoelectron. **9**, 67–70 (2006)
24. V. Balakrishnan, H.-P. Phan, T. Dinh, D.V. Dao, N.-T. Nguyen, Thermal flow sensors for harsh environments. Sensors **17**, 2061 (2017)
25. V. Balakrishnan, T. Dinh, H.-P. Phan, D.V. Dao, N.-T. Nguyen, Highly sensitive 3C-SiC on glass based thermal flow sensor realized using MEMS technology. Sens. Actuators A Phys. (2018)
26. T. Dinh, H.-P. Phan, A. Qamar, P. Woodfield, N.-T. Nguyen, D.V. Dao, Thermoresistive effect for advanced thermal sensors: fundamentals, design considerations, and applications. J. Microelectromech. Syst. **26**, 966–986 (2017)
27. E. Meng, P.-Y. Li, Y.-C. Tai, A biocompatible Parylene thermal flow sensing array. Sens. Actuators A Phys. **144**, 18–28 (2008)
28. C. Chen, Evaluation of resistance–temperature calibration equations for NTC thermistors. Measurement **42**, 1103–1111 (2009)
29. S.M. Sze, K.K. Ng, *Physics of Semiconductor Devices* (John Wiley & Sons, Hoboken, 2006)
30. P. Yih, J. Li, A. Steckl, SiC/Si heterojunction diodes fabricated by self-selective and by blanket rapid thermal chemical vapor deposition. IEEE Trans. Electron Devices **41**, 281–287 (1994)
31. N. Zhang, C.-M. Lin, D.G. Senesky, A.P. Pisano, Temperature sensor based on 4H-silicon carbide pn diode operational from 20 °C to 600 °C. Appl. Phys. Lett. **104**, 073504 (2014)

32. S. Rao, G. Pangallo, F. Pezzimenti, F.G. Della Corte, High-performance temperature sensor based on 4H–SiC Schottky diodes. IEEE Electron Device Lett. **36**, 720–722 (2015)
33. S. Rao, G. Pangallo, F.G. Della Corte, Highly linear temperature sensor based on 4H-silicon carbide pin diodes. IEEE Electron Device Lett. **36**, 1205–1208 (2015)
34. S. Rao, G. Pangallo, F.G. Della Corte, 4H–SiC pin diode as highly linear temperature sensor. IEEE Trans. Electron Devices **63**, 414–418 (2016)
35. S.B. Hou, P.E. Hellström, C.M. Zetterling, M. Östling, 4H–SiC PIN diode as high temperature multifunction sensor. Mater. Sci. Forum 630–633 (2017)
36. T. Dinh, H.-P. Phan, N. Kashaninejad, T.-K. Nguyen, D.V. Dao, N.-T. Nguyen, An on-chip SiC MEMS device with integrated heating, sensing and microfluidic cooling systems. Adv. Mater. Interfaces **1**, 1 (2018)
37. T. Dinh, H.-P. Phan, D.V. Dao, P. Woodfield, A. Qamar, N.-T. Nguyen, Graphite on paper as material for sensitive thermoresistive sensors. J. Mater. Chem. C **3**, 8776–8779 (2015)
38. T. Dinh, H.-P. Phan, T. Kozeki, A. Qamar, T. Fujii, T. Namazu et al., High thermosensitivity of silicon nanowires induced by amorphization. Mater. Lett. **177**, 80–84 (2016)
39. T. Neda, K. Nakamura, T. Takumi, A polysilicon flow sensor for gas flow meters. Sens. Actuators A Phys. **54**, 626–631 (1996)
40. C. Lyons, A. Friedberger, W. Welser, G. Muller, G. Krotz, R. Kassing, A high-speed mass flow sensor with heated silicon carbide bridges, in *Proceedings MEMS 98. The Eleventh Annual International Workshop on Micro Electro Mechanical Systems, 1998*, 1998, pp. 356–360
41. R. Ahrens, K. Schlote-Holubek, A micro flow sensor from a polymer for gases and liquids. J. Micromech. Microeng. **19**, 074006 (2009)
42. J.-G. Lee, M.I. Lei, S.-P. Lee, S. Rajgopal, M. Mehregany, Micro flow sensor using polycrystalline silicon carbide. J. Sens. Sci. Technol. **18**, 147–153 (2009)
43. P. Bruschi, M. Dei, M. Piotto, A low-power 2-D wind sensor based on integrated flow meters. IEEE Sens. J. **9**, 1688–1696 (2009)
44. M.I. Lei, *Silicon Carbide High Temperature Thermoelectric Flow Sensor* (Case Western Reserve University, 2011)
45. S. Issa, H. Sturm, W. Lang, Modeling of the response time of thermal flow sensors. Micromachines **2**, 385–393 (2011)
46. C. Sosna, T. Walter, W. Lang, Response time of thermal flow sensors with air as fluid. Sens. Actuators A Phys. **172**, 15–20 (2011)
47. H. Berthet, J. Jundt, J. Durivault, B. Mercier, D. Angelescu, Time-of-flight thermal flowrate sensor for lab-on-chip applications. Lab Chip **11**, 215–223 (2011)
48. A.S. Cubukcu, E. Zernickel, U. Buerklin, G.A. Urban, A 2D thermal flow sensor with sub-mW power consumption. Sens. Actuators A Phys. **163**, 449–456 (2010)
49. X. She, A.Q. Huang, Ó. Lucía, B. Ozpineci, Review of silicon carbide power devices and their applications. IEEE Trans. Ind. Electron. **64**, 8193–8205 (2017)
50. J.W. Judy, Microelectromechanical systems (MEMS): fabrication, design and applications. Smart Mater. Struct. **10**, 1115 (2001)
51. F. Mailly, A. Martinez, A. Giani, F. Pascal-Delannoy, A. Boyer, Design of a micromachined thermal accelerometer: thermal simulation and experimental results. Microelectron. J. **34**, 275–280 (2003)
52. G.M. Rebeiz, *RF MEMS: Theory, Design, and Technology* (John Wiley & Sons, Hoboken, 2004)
53. S.-H. Tsang, A.H. Ma, K.S. Karim, A. Parameswaran, A.M. Leung, Monolithically fabricated polymermems 3-axis thermal accelerometers designed for automated wirebonder assembly, in *IEEE 21st International Conference on Micro Electro Mechanical Systems, 2008. MEMS 2008*, 2008, pp. 880–883
54. T.X.D. Van Thanh Dau, D.V. Dao, S. Sugiyama, Design and simulation of a novel 3-DOF Mems Convective Gyroscope. IEEJ Trans. Sens. Micromach. **128**, 219–224 (2008)
55. P.R. Gray, P.J. Hurst, R.G. Meyer, S.H. Lewis, *Analysis and Design of Analog Integrated Circuits* (John Wiley & Sons, New York, 2008)

56. A. Singh, Film thickness and grain size diameter dependence on temperature coefficient of resistance of thin metal films. J. Appl. Phys. **45**, 1908–1909 (1974)
57. F. Lacy, Developing a theoretical relationship between electrical resistivity, temperature, and film thickness for conductors. Nanoscale Res. Lett. **6**, 1 (2011)
58. F. Lacy, Using nanometer platinum films as temperature sensors (constraints from experimental, mathematical, and finite-element analysis). IEEE Sens. J. **9**, 1111–1117 (2009)
59. A. Feteira, Negative temperature coefficient resistance (NTCR) ceramic thermistors: an industrial perspective. J. Am. Ceram. Soc. **92**, 967–983 (2009)

Chapter 4
Fabrication of SiC MEMS Sensors

Abstract This chapter describes the approaches for the growth of high-quality silicon carbide. The doping methods for n-type and p-type SiC with the inclusion of selective doping are presented. Various fabrication strategies are introduced, including wet etching and oxidation. Fundamental properties of Ohmic and Schottky contacts to SiC are included. This chapter also gives brief examples of fabrication processes to achieve standard MEMS structures such as cantilevers and membranes.

Keywords Deep reactive-ion etching (DRIE)
Chemical vapour deposition (CVD) · Ohmic contact · Schottky contact

4.1 Growth and Doping

4.1.1 Growth of SiC

The growth of bulk single crystal hexagonal SiC (e.g. 4H-SiC and 6H-SiC) wafers was successfully conducted by seeded sublimation [1, 2]. For example, 4H-SiC can be grown from seed crystals using a high-temperature chemical vapour deposition (DTCVD) process. The physical vapour transport was well established for the growth of 4H- and 6H-SiC bulk crystals [3–5]. Though commercialised by Cree Inc. since 1990s, the cost of these hexagonal wafers is quite high as it requires high temperatures (e.g. 1,500–2,400 °C) for a high growth rate [6]. 4H- and 6H-SiC have been widely employed for commercialised high power electronics. However, the use of hexagonal SiC has been very limited for MEMS sensors due to the difficulty of subsequent fabrication processes such as etching and releasing suspended structures [7–9].

The growth of cubic silicon carbide (3C-SiC) on large-area Si substrates is of much interest for MEMS sensors as it is an alternative solution to reduce the cost of such SiC wafers and it offers easy fabrication of MEMS structures [10–17]. There are two main approaches for the growth of 3C-SiC, including physical and chemical depositions. The physical depositions of SiC films typically refer to the sputtering

© The Author(s) 2018
T. Dinh et al., *Thermoelectric Effect in SiC for High-Temperature
MEMS Sensors*, SpringerBriefs in Applied Sciences and Technology,
https://doi.org/10.1007/978-981-13-2571-7_4

method in which the plasma bombards a SiC target and SiC materials are deposited on a substrate.

Chemical vapour deposition (CVD) methods are used for growth of high crystallinity and high-quality films. CVD approach for growth of SiC refers to a decomposition of reactant gases (e.g. silane SiH_4 and C_3H_6) and formation of SiC on a substrate in a high-temperature chamber. The first technique of CVD growth of SiC is atmospheric pressure chemical vapour deposition (APVCD). In this method, the Si substrate is cleaned using a carbonisation process before a growth of SiC is performed at a high temperature of 1,300 °C using Si and C precursors [18, 19]. This approach offers high growth flow rates of epitaxial SiC films with both n-type and p-type dopants. Low-pressure chemical vapour deposition (LPCVD) is a second CVD method employing a low chamber pressure with much lower growth rate (<1 nm/cycle) and more diversity of precursors. This method generates high-quality films with better thickness uniformity (>99% across 150 mm wafer) and smoothness (<5 nm RMS roughness) [20]. Consequently, LPCVD has become a popular approach to grow SiC films on different substrates (e.g. Si, SiO_2, Si_3N_4) for MEMS applications. On the other hand, plasma-enhanced chemical vapour deposition (PECVD) requires much lower deposition temperatures (e.g. from 200 to 400 °C) [21, 22]; hence, it is suitable for IC MEMS processing and coating/protecting of MEMS components. This approach typically produces amorphous SiC and requires post-annealing processes for recrystallisation of SiC. The pros and cons of these CVD approaches are summarised in Table 4.1. Figure 4.1 shows the schematic sketch of concepts of SiC growth method using CVD.

Table 4.1 Chemical vapour deposition method for growth of SiC films

Approach	Temperature (°C)	Pressure	Growth rate	Precursors
APCVD	1,300	Atmosphere	High	Limited
LPCVD	1,000	Low pressure	Low	Diversity
PECVD	200–400	–	–	Diversity

Fig. 4.1 Chemical vapour deposition (CVD) process

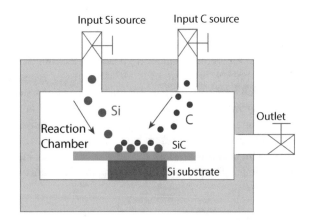

Table 4.2 Doping substances for epitaxial doping of SiC films

Doping types	Lely process	Liquid-phase epitaxy (LPE)	Chemical vapour deposition (CVD)
n-type doping	N_2, P	N_2	N_2, NH_3, PH_3, PCl_3
p-type doping	Ga, B and Al	Al	$Al(CH_3)_3$, $Al(C_2H_5)_3$, B_2H_6, BBr_3, $AlCl_3$

4.1.2 Doping of SiC

Doping in SiC MEMS devices has typically performed in the epitaxial growth process (so call epitaxial doping) and after the growth process using ion implantation (selective doping) [23, 24]. Since SiC has very low diffusion coefficient at a temperature of below 1,800 °C because of its high strength bond, it is challenging to implant ions to dope SiC. Especially, ion implantation requires high-temperature implant and subsequent high-temperature annealing (500–1,700 °C) to activate dopants. This approach leads to the damage of SiC lattice and results in amorphised materials and high defect density at the room temperature. However, ion implantation has its own advantage in terms of selective doping which is required for logic structures and embedded sensing elements.

Expitaxial doping offers a wide range of doping level from 10^{14} to 10^{19} cm^{-3} with two common types of dopants (i.e. n-type with nitrogen and p-type with aluminium). Table 4.2 shows different types of doping substances for epitaxial doping of SiC.

In the epitaxial doping process, Si/C source gas ratio is adjusted to control the amount of dopants substituting into SiC lattice sites. The epitaxial doping mechanism is based on the replacement of N on C sites for n-type SiC and Al on Si sites for p-type SiC. This technique has also employed to dope boron and phosphorous into the SiC lattice. In general, p-type doping faces more challenging than n-type doping and requires high temperatures to activate the acceptors over the required energy (ionisation energy). Epitaxial doping has its own advantage in terms of no lattice damage, and therefore, it eliminates the formation of defects and the requirement of high-temperature annealing.

4.2 Etching of SiC

Etching techniques of SiC can be divided into dry etching and wet etching, where the latter can be used for defect determination owing to its low cost and simplicity. The etching rate of defect-free SiC and local defect area is different with etchants and etching conditions; hence, defect-selective etching has been used. Wet etching has been typically employed as an isotropic etching method. Different to wet etching, dry etching can be highly anisotropic to fabricate high aspect ratio structures. This

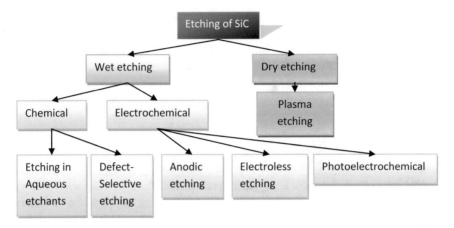

Fig. 4.2 Etching categories of SiC [25, 26]

method has a low etching selectivity to different materials, and it can cause the surface damage due to ion bombardment.

Wet etching of SiC refers to the oxidation of SiC surface and subsequent dissolution of silicon carbide oxides. This method can be categorised into two groups of chemical (conventional) etching and electrochemical etching. Figure 4.2 shows an overall picture of SiC etching approaches [25, 27–29]. In anodic wet etching, SiC is oxidised by providing a bias voltage to replace bonding electrons by holes, while it is put inside an electrolyte (e.g. KOH) to dissolve the resulting oxides. Similar to anodic etching, electroless etching requires an external voltage to deplete electrons in the valence band. The oxidation process is driven by oxidising agent in the electrolytes. If an illumination source with energy not less than energy gap E_g is used, the photogenerated holes will assist the oxidation of SiC, while the photogenerated electrons are consumed by reducing oxidizing agent. This method is known as photoassisted electroless etching. It is known as photoelectrochemical etching (required electrical contact and counter electrodes) if photogenerated electrons are consumed by reducing reaction on the counter electrodes.

Chemical etching employs the reactive molecules of etchants to breakdown the SiC chemical bonds, resulting in the formation of oxides. The oxides are subsequently dissolved in the etchants [25]. This method is not involved in electrolytes and external voltage. However, due to the chemical inertness of SiC, it is challenging to chemically etch SiC in conventional aqueous etching solutions. Chemical etching of SiC has been successfully demonstrated with phosphoric acid (heated to 215 °C) and alkaline solution $K_3Fe(CN)_6$ (heated above 100 °C), but suffering very low etching rate or impractical. Single-crystalline SiC typically need be amorphised before chemically etched. The amorphisation has been done with implantation of high dose ions (e.g. Xe^+), and then, chemical etching was performed in acid solutions (e.g. HF:HNO$_3$ 1:1) to achieve smooth surface. Due to the difficulty of chemical etching of SiC, electrochemical etching is more popular and will be discussed in the next section.

4.2.1 Electrochemical Etching

The electrochemical etching involves in oxidation processes with silicon compound products such as SiO_x and CO_x, presented in the following reactions [30–34]:

$$SiC + 4H_2O + 8h^+ \rightarrow SiO_2 + CO_2 + 8H^+ \tag{4.1}$$

$$SiC + 2H_2O + 8h^+ \rightarrow SiO + CO + 4H^+ \tag{4.2}$$

$$SiO_2 + 6HF + 2H^+ \rightarrow SiF_6^{2-} + 2H_2O \tag{4.3}$$

Several electrolytes used include but are not limited to KOH, NaOH and HF. While β-SiC (3C-SiC) can be directly etched by electrochemical method, electrochemical etching of α-SiC (4H-SiC and 6H-SiC) typically results in amorphous SiC which requires a thermal oxidation process before etched. The etching of n-type 3C-SiC in HF was performed in HF solution with an assistance of light illuminations. UV light-assisted HF etching of SiC was proven to show a much higher etching rate in comparison to the use of green lights (e.g. 100 times higher). However, photoassisted etching of α-SiC typically results in a formation of oxidation layers and requires further steps to etch these layers.

Anodizing etching of α-SiC in HF solution usually forms a porous SiC layer. This porous layer is thermally oxidised following by an oxide etching process. Dopant-type selective etching is also demonstrated with the etching of n-type over p-type SiC in p-n SiC junctions with an etching selectivity of 10^5. Etching of p-type SiC over n-type SiC has typically performed using anodic etching in dark conditions.

4.2.2 Chemical Etching

In chemical etching, defect-selective etching is preferable for the determination of defects in SiC crystals such as dislocations, nanopipes as well as inversion domains [35–37]. Research on revealing the dislocations in SiC crystals attracted great attention as the planer defects and stacking faults affect the electrical and optical properties of SiC. The etching of SiC was conventionally performed in various molten salts including $KClO_3$, K_2CO_3, K_2SO_4 and KNO_3. However, the etching is unstable and requires high temperatures of 900–1,000 °C. Etching of KOH and its mixtures with other salts can reduce the etching temperatures to 300–600 °C. An important note for etching of SiC with these molten salts is the compulsory presence of oxygen from their decomposition (Na_2O_2) or surrounding oxygen. The presence of oxygen can increase the etching rate to 2.5–5 times. These suggested that the surface oxidation is involved in the etching process.

4.2.3 Dry Etching or Reactive-Ion Etching (RIE)

As wet etching requires high-temperature conditions or complicated photoassistance, plasma-based (dry) etching has played an important role for the fabrication of SiC MEMS devices [38–40]. For dry etching of SiC, mixtures of fluorinated gases and oxygen are used to remove Si and C atoms based on the following reactions [28, 29]:

$$Si + mF \rightarrow SiF_m \quad (m = 1\text{--}4), \tag{4.4}$$

$$C + mF \rightarrow CF_m, \tag{4.5}$$

$$C + nO \rightarrow CO_n \quad (n = 1\text{--}2), \tag{4.6}$$

The combination of reaction is as follows:

$$SiC + mF + nO \rightarrow SiF_m + CO_n + CF_m. \tag{4.7}$$

When a low oxygen percentage (5–20%) is added to the fluorinated gases, the etching rate achieved the highest value. In addition, RIE of SiC can be performed in mixtures of fluorinated oxygen gases including CHF_3/O_2, $CBrF_3/O_2$, $CF4/O_2$, SF_6/O_2, NF_3/O_2. RIE etching of SiC is also demonstrated in fluorinated mixture gases such as CF_4/CHF_3, SF_6/CHF_3, and NF_3/CHF_3 and SF_6/NF_3. Table 4.3 provides a few etching conditions with different types of source gases with the etching rate. The RIE etching of SiC shows the unstable etching rates from few nm to 1 μm per minute. The etching rate increases with increasing input rf power.

4.3 Ohmic and Schottky Contacts to SiC

4.3.1 Ohmic Contact

An Ohmic contact refers to a junction between a metal electrode and a semiconductor, without limiting the current flow through. Figure 4.3 shows the formation of an Ohmic contact of a metal to an n-type semiconductor. As the work function Phi_s of the semiconductor is larger than that of metal phi_m, electrons from metal will tunnel into the conduction band of semiconductor until the equilibrium prevents the movement of electrons. This leads to the uniformity of Femi level across the system. The accumulation region is formed with excess electrons near the junction of the metal and the semiconductor. There are electrons in both sides of the junction, resulting in no barrier to prevent the movement of electrons across the junction under the applied electric field. Since the accumulation region and metal have a higher electron density compared to the semiconductor, it has a smaller resistance and a smaller voltage drop when a voltage is applied to the system.

Table 4.3 Reactive-ion etching (RIE) of 3C-SiC, 4H-SiC and 6H-SiC [28]

Polytypes	Process type	Source gas	Etching conditions (pressure, power, dc bias, flow rate)	Etching rate (nm/min)
3C-SiC	RIE (rf)	CF_4/O_2	180–200 mT, 0.8 W/cm^2, 67% O_2, 33% CF_4	6–26
4H-SiC, 6H-SiC	RIE (rf)	SF_6	20 mT, 250 W, −220 to −250 V, 20, 35 sccm	49, 42, 57, 53
6H-SiC	RIE (rf)	SF_6/O_2 NF_3/O_2	20 mT, 200 W, −220 to −250 V, SF_6: O2 = 18: 2 (sccm) NF_3: O_2 = 18: 2 (sccm)	45, 57
6H-SiC	RIE (rf)	SF_6/O_2	50 mT, 200 W, −250 V, SF_6: O_2 = 5: 5 (sccm)	36
6H-SiC	RIE (rf)	SF_6/O_2 CF_4/O_2 with N_2 additive	190 mT, 300 W, CF_4: O_2: N_2 = 40: 15: 10 (sccm), SF_6: O_2: N_2 = 40: 2: 0 (sccm)	220, 300
4H-SiC, 6H-SiC	RIE (rf)	NF_3	20 mT, 250 W, −220 to −250 V, 20, 35 sccm	56.5, 54.0, 63.0
4H-SiC, 6H-SiC	RIE (rf)	NF_3	225 mT, 275 W, −25 to −50 V, 95–110 sccm	150
6H-SiC	RIE (rf)	$Cl_2/SiCl_4/O_2$ and Ar/N_2	190 mT, 300 W, Cl_2: $SiCl_4$: O_2: N_2 = 40: 20: 8: 10 (sccm), Cl_2: $SiCl_4$: O_2: Ar =40: 20: 0: 10 (sccm)	160, 190
3C-SiC, 6H-SiC	RIE (microwave)	SF_6/O_2	1 mT, 1,200 W, −20 to −110 V, SF_6: O_2 = 4: 0–8 (sccm), SF_6: O_2 = 4: 0–6 (sccm)	100– 270
4H-SiC, 6H-SiC	RIE (microwave)	CF_4/O_2	1 mT, 650 W, −100 V, CF_4: O_2 =41.5: 8.5 (sccm)	70

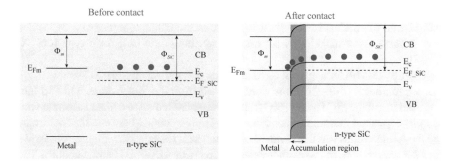

Fig. 4.3 Formation concept of Ohmic contact in SiC

The Ohmic contact with SiC is crucial to develop SiC MEMS sensors for harsh environments. The degradation of metal contact and the large specific contact resistance is one of the great challenges for high-temperature SiC sensors and high-power

SiC electronics. The degradation of meta/SiC contact comes from the formation of oxides and silicides during the long-time service at high temperatures. Therefore, multiple stacks of metal layers have been employed to maintain the low specific contact resistance, excellent resistance to oxidation, as well as stable contact. To form the Ohmic contact, the following potential barrier between a metal and SiC should be as low as possible [41, 42]:

$$q\phi_B = q\phi_m - \chi_s, \tag{4.8}$$

where $q\phi_B$ and χ_s are the work function of the metal and the electron affinity. However, the work function of almost all metals lies between 4.5 and 6 eV, while the electron affinity of n-type and p-type SiC is 4 and 7 eV, respectively [43–45]. This leads to the challenges of forming Ohmic contact with low contact resistance to SiC. Therefore, the high-temperature annealing conditions are crucial to form an Ohmic contact with SiC. The typical metals used for SiC Ohmic contact are Ti and Ni to achieve a low contact resistance of 10^{-4}–10^{-6} Ω cm^2 [46–51].

Ohmic contacts to n-type SiC is typically formed with high-temperature annealing from 800 to 1,100 °C. In these conditions, it is hard to reveal the Ohmic formation mechanism due to the complicated interaction/interdiffusion at the interfaces. Some hypothesis has been given to the formation of silicides/carbides including Ni$_2$Si and TiSi$_2$ at high temperatures, or the diffusion of C atoms from SiC [46]. These C atoms act as donors to reduce the barrier height to form an Ohmic contact.

For p-type SiC, Ti/Al and Ni/Ti/Al stacks have been typically used [52–54]. The formation of Al$_3$C$_4$ and Ti$_3$SiC$_2$ reduces the Schottky barrier height of metal/SiC. Some studies have demonstrated the Ohmic contact with p-type SiC with Al-based contacts where the diffusion of Al at high temperatures results in a formation of a heavily doped layer and reduces the contact resistance. A recent study has demonstrated an Ohmic contact. Table 4.4 shows some recent progress to achieve the Ohmic contact with n-type 4H-SiC and 6H-SiC [42]. In addition, Table 4.5 shows the different annealing conditions and parameters to form the Ohmic contact with p-type 4H-SiC and 6H-SiC. It is important to note that the Ohmic contact with 3C-SiC is easy to form using nickel (Ni) for n-type and aluminium (Al) for p-type [16, 51, 55–57].

Recently, 3C-SiC film grown on Si has been transferred on a glass substrate using focused ion beam [58]. SiC films were released from the Si substrate, and FIB tool was used to cut the SiC frames. SiC frames were handled on to a glass substrate with two pre-deposited aluminium contacts. The Ohmic contact between the SiC frame and the Al electrodes was then formed using tungsten for adhesion. Figure 4.4a shows the schematic sketch of the as-fabricated device with Al/W/SiC Ohmic contact. The scanning electron microscopy (SEM) image of this platform is presented in Fig. 4.4b. The Ohmic contact was confirmed by performing the current–voltage measurements at different temperatures as shown in Fig. 4.4c. The linear I–V characteristics for every temperature indicate the excellent Ohmic contact formed between SiC and Al/W contacts.

Table 4.4 Ohmic contact with n-type 4H-SiC and 6H-SiC [42]

Metallisation stack	SiC polytypes	Doping level $(10^{18}$ cm$^{-3})$	Temperature annealing (°C)	Time (min)	Condition	Sheet resistance $(10^{-5}$ Ω cm$^2)$
Ni	6H-SiC	0.0055	1,070	10	Vacuum	>800
Ni/Ti	6H-SiC	0.45	960	10	Vacuum	500
C	4H-SiC	13	1,050	30	Ar	431
Ni	6H-SiC	1	1,000	5	Ar	300
Ti/Sb/Ti	6H-SiC	0.35	960	10	Vacuum	~270
Ti	6H-SiC	0.35	960	10	Vacuum	~250
Ti	4H-SiC	No annealing				225
Ni/Ti	6H-SiC	0.47	1,065	10	Vacuum	105
Pt/Si	6H-SiC	0.35	1,065	15	Vacuum	100
Ni	6H-SiC	0.55	960	10	Vacuum	80
Si	6H-SiC	0.55	960	10	Vacuum	80
Ti/Sb/Ti	6H-SiC	2.3	960	10	Vacuum	~77
Ti/Al/Ti	6H-SiC	2.3	1,065	10	Vacuum	73.6
Ni/Si	6H-SiC	15	300	540	N_2	69
Ti/TiN/Pt/Ti/Ti	4H-SiC	30	1,050	N.R.	Ar	65
Ti	6H-SiC	2.3	960	10	Vacuum	~55
Ti$_3$SiC$_2$	4H-SiC	15	950	1	Ar	50
Si/WNi	4H-SiC	5–7	1,100	60	Ar	50
C	4H-SiC	13	1,350	30	Ar	43
Ni	6H-SiC	1.7	960	10	Vacuum	42
TiW	4H-SiC	50	950	5	Ar	40.8
Pt	4H-SiC	4.2	1,150	15	Vacuum	40
Au/TaSiN/Ni/Si	4H-SiC	N.R.	1,000	3	N_2	~40
Au/Ni/Si	4H-SiC	N.R.	1,000	3	N_2	~40
Pt/TaSi$_x$/Ni	4H-SiC	11	950	30	Ar	35
Pt/TaSi$_x$/Ni	4H-SiC	0.01	900	5	Ar	34
Ni/Ti	6H-SiC	1.7	970	10	Vacuum	33
Ta/Ni/Ta	6H-SiC	0.6	800	10	Ar	30
WSi$_2$	6H-SiC	>10	1,000	20	H_2	24
Ni/Ti	6H-SiC	19	960	10	Vacuum	22
Ti/TiN/Pt/Ti/Ni	4H-SiC	30	950	N.R.	Ar	21
Ti	4H-SiC	1	400	5	N_2	20.7
Co/C	4H-SiC	1.6	800	120	Vacuum	20.4
Al/Ti/N	4H-SiC	10	800	30	UHV	20
Al/Ni	4H-SiC	13	1,000	5	UHV	18
Ti/InN	4H-SiC	9.8	No annealing			18
Pt/TaSi$_2$/Ti	6H-SiC	7	600	30	N2	16.8
Nb	4H-SiC	3	1,000	10	Vacuum	15.3
Ni	6H-SiC	19	960	10	Vacuum	15
Ni/C	4H-SiC	1.6	800	120	Vacuum	14.3

Table 4.5 Ohmic contact with p-type 4H-SiC and 6H-SiC [42]

Metallisation stack	SiC polytypes	Doping level (10^{18} cm^{-3})	Temperature annealing (°C)	Time (min)	Condition	Sheet resistance (10^{-5} Ω cm^2)
Ni/Al	4H-SiC	7.2	1,000	5	UHV	1,200
Ni/Ti/Al	4H-SiC	4.5	800	30	UHV	220
Ni/Ti	4H-SiC	~100	950	~1	N$_2$	130
Ni	4H-SiC	100	1,000	1	N$_2$	~100
Au/NiAl/	4H-SiC	>100	600	30	Vacuum	~80
Au/NiAl/Ti	4H-SiC	>100	600	30	Vacuum	~50
Al	4H-SiC	4.8	1,000	2	Vacuum	42
Al	4H-SiC	4.8	1,000	2	Vacuum	42
Ti/Al	6H-SiC	16	900	4	N$_2$	40
Co/Si/Ti	4H-SiC	3.9	850	1	Vacuum	40
Pt/Si	6H-SiC	7	1,100	5	Vacuum	28.9
Al/Ti/Al	4H-SiC	4.8	1,000	2	Vacuum	25
Au/Pt-N/TaSiN/Al70Ti30	4H-SiC	~1,000	1,000	2	Vacuum	~20
Si/Al/Ti	4H-SiC	24	1,020	2.5	Ar	17
Pt	4H-SiC	10	1,100	5	Vacuum	~15
Al/Ti/Ge	4H-SiC	4.5	600	30	Vacuum	10.3
AlSiTi	4H-SiC	30–50	950	7	Ar	9.6
Au/CrB2	6H-SiC	13	1,100	15	Vacuum	9.58
Ni/Pt/Ti/Al	4H-SiC	6–8	1,000	2	Vacuum	9
Al/Ti/Ni	4H-SiC	4.5	800	30	Vacuum	7
Au/Pt-N/TaSiN/Ni (7% V)	4H-SiC	~1,000	900	1	Vacuum	~8
Ti/Al	4H-SiC	10	1,000	2	Vacuum	~7
Al/W	4H-SiC	100	850	1	Ar	6.8
Au/Sn/Pt/TaRuN/Ni/Al	4H-SiC	1,000	N.R.	N.R.	N.R.	2
W/NiAl/Ti	4H-SiC	>100	975	2	Ar	~5.5
Pd	4H-SiC	50	700	5	N$_2$	5.5
Al/Ti	4H-SiC	100	850	1	Ar	4.8
Pt/Si	4H-SiC	10	1,000	5	Vacuum	~4.4
Au/Pd	4H-SiC	30–50	850	15	Ar	4.19
Au/Pd/Al	4H-SiC	30–50	900	5	Ar	4.08
W/Ni/Al	N.R.	10	850	2	N.R.	4
Ni	4H-SiC	100	1,000	1	Ar	~3
Au/Pd/Ti/Pd	4H-SiC	30–50	900	N.R.	Ar+1%H$_2$	2.9
Au/Ta/TaRu/Ni	4H-SiC	~1,000	850	1	Ar	~2
TiC	4H-SiC	20	500	3	Ar+10% H$_2$	2
Ti/Al	4H-SiC	4.5	1,000	2	Vacuum	2
Au/TiW/Ti/Pd	4H-SiC	100	400+850	1.5+1	N$_2$	1.6
Au/Ti/Al	4H-SiC	30–50	900	5	Ar+1% H$_2$	1.42
Al–Ti	N.R.	13	1,000	2	Vacuum	1.1
Au/Ru/TaRu/Ni/Al	4H-SiC	~1,000	850	1	Ar	~1
Al/Ti/Al	4H-SiC	10	1,050	10	N.R.	0.5
Al/Ti	4H-SiC	10	1,050	N.R.	Ar	0.1

Fig. 4.4 Ohmic contact with p-type 3C-SiC using tungsten (W) and aluminium (Al). Reprinted with permission from Ref. [58]

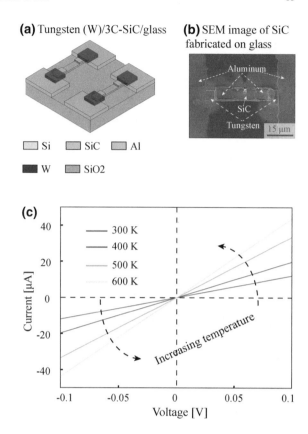

(a) Tungsten (W)/3C-SiC/glass

(b) SEM image of SiC fabricated on glass

4.3.2 Schottky Contact

Similar to the formation of the Ohmic contact, the formation of a Schottky contact with SiC is shown in Fig. 4.5. The thermal equilibrium will bring the electrons from the semiconductors to the metal until the Fermi level is uniform across the system, creating a barrier Φ_B, which is called a Schottky barrier height. The current density J passing through this junction can be described as follows:

$$J = J_o\left[\exp\left(\frac{eV}{kT}\right) - 1\right] \tag{4.9}$$

where V is the external electric field and n is the ideality factor. Figure 4.6a shows a structure of a Schottky diode of Ti/Al stacks to n-4H-SiC which has been used to develop high-performance temperature sensors [52, 61]. Figure 4.6b illustrates the current–voltage characteristics of the Schottky barrier under elevated temperatures. The sensitivity of the Schottky sensor was found to be 5.11 mV/K with good linearity

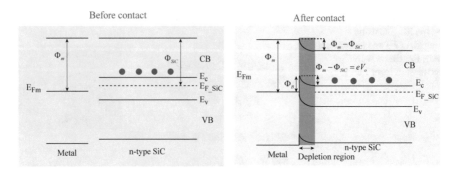

Fig. 4.5 Formation of Schottky contact between n-type SiC and a metal [59, 60]

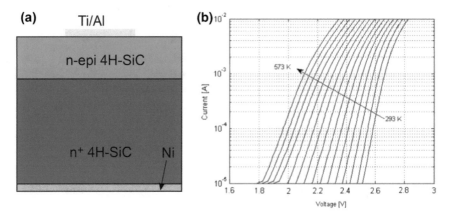

Fig. 4.6 a Structure of a Schottky diode formed between n-type 4H-SiC and Ti/Al metal contacts. **b** Current–voltage characteristics of the SiC Schottky diode. Reprinted with permission from Ref. [61]

up to 300 °C and excellent long-term stability, which is one of the best performances for the SiC Schottky temperature sensors to date.

In the Schottky temperature sensors, the Schottky barrier and the ideality factor can be calculated from the current–voltage measurement. Recent studies have demonstrated an excellent stability of the Schottky barrier and ideality factor under different elevated temperature conditions. Figure 4.7 shows a slight increase in the ideality factor with increasing temperature, and a Schottky barrier of approximately 1.6 eV [62].

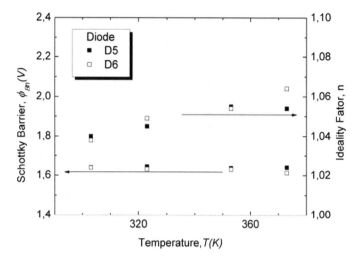

Fig. 4.7 Dependence of the Schottky barrier and ideality factor on temperature. Reprinted with permission from Ref. [62]

4.4 Fabrication Processes of SiC MEMS Sensors

4.4.1 Surface Micromachining

Figure 4.8 shows the standard example of surface micromachining technique for fabrication of MEMS SiC devices. The first process is to deposit an insulation layer (e.g. silicon dioxide and silicon nitride) on top of the Si substrate. The same thermal expansion of the insulation layer with SiC such as silicon nitride will be preferable as it is suitable for high-temperature applications. The next step involves the deposition of the poly-Si, or the oxide is used as a sacrificial layer. The SiC layer is then grown on the poly-Si layer. Figure 4.8a shows the structure of the platform.

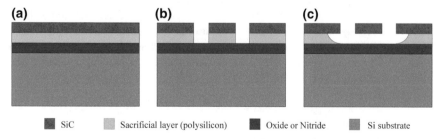

Fig. 4.8 Surface micromachining technology for SiC MEMS devices. **a** Growth of SiC/polysilicon/oxide(nitride)/Si. **b** Patterning of SiC and sacrificial layer. **c** Etching of the sacrificial layer to form the released SiC structures

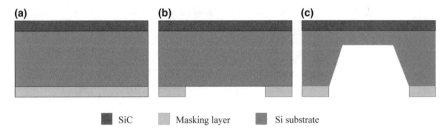

Fig. 4.9 Bulk-micromachining technology for SiC MEMS devices

The utilisation of the poly-Si as the sacrificial layer leads to the ease of etching and patterning with KOH or TMAH to release the SiC microstructures. Therefore, the oxide functions as a protective layer for the underlying silicon as the final etching process is to remove the sacrificial area. In the cause of using oxide as a sacrificial layer, HF-based solutions can be employed. Since SiC is highly resistant to HF solutions, no protection of the functional layer is required. Figure 4.8b illustrates the results of patterning SiC and the sacrificial layer. Figure 4.8c shows the etching step for releasing the SiC structures.

4.4.2 Bulk Micromachining

Figure 4.9 shows a simple process for bulk micromachining of SiC devices such as pressure sensors.

Figure 4.9a illustrates the starting structure for making the membrane from the back of the wafer. The process (Fig. 4.9b, c) is similar to fabrication strategy applied for silicon-on-insulator (SOI) wafers. The only difference is the high resistance of SiC to etchants, which shows the function of SiC as a stop-etch barrier. In addition, etching from the front side can also be employed to fabricate freestanding SiC structures, in which SiC is etched first, followed by the bulk etching of silicon substrates. However, bulk etching of 4H-SiC and 6H-SiC is much more difficult than that of 3C-SiC on silicon as etching rate of SiC is very slow when using the conventional wet etching methods. Recently, a novel fabrication process has been proposed to etch ultrathin 4H-SiC thin membranes. The process employs a fast reactive-ion etching (RIE) method, followed by a dopant selective reactive-ion etching (DSRIE) approach. This process can produce silicon carbide (SiC) diaphragms with a thickness of less than 10 μm [63].

A recent work has demonstrated a scribe laser process for bulk machining of a pressure sensor [9], which is illustrated in Fig. 4.10. Figure 4.10a shows the four main steps of the fabrication process of the pressure sensor. The fabrication process starts from a 4H-SiC wafer consisting of a p-type top layer, an n-type 4H-SiC and a thick 4H-SiC substrate. The piezoresistors were formed on top of the wafer in the (0001) face. The piezoresistors were designed with a U-shaped configuration using

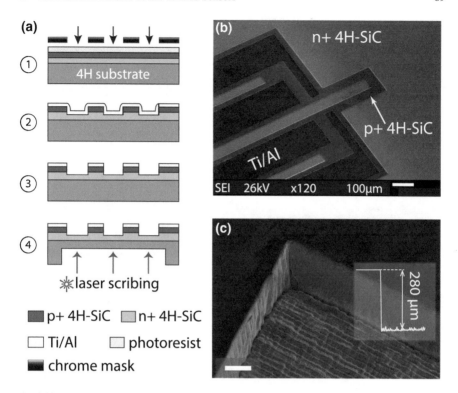

Fig. 4.10 Bulk-micromachining technology for 4H-SiC MEMS pressure sensor using a laser scribing process [9]. **a** Fabrication process. **b** Scanning electron microscopy (SEM) image of 4H-SiC piezoresistor fabricated on top of a 4H-SiC wafer. **c** SEM image of the back of the pressure sensor, formed by a UV laser scribing. Reprinted with permission from Ref. [9]

the functional p-type layer. The photolithography process was employed in step 1 to pattern the photoresist layer which has the same shape as the piezoresistors. In the second step, an inductive coupled plasma (ICP) etching process was performed (step 2 using an etcher) to form the 4H-SiC piezoresistors. The etching time was calculated to overetch the p-type layer to ensure the isolation of the piezoresistors from the n-type substrate (Fig. 4.10a). The ICP etching rate was estimated to be about ~100 nm/min. In step 3, a Ti/Al metallisation stack was formed on the top of the p-type layer and then patterned using a photolithography and an etching process. After that, the Ti/Al metal layers were annealed at 1,000 °C in N_2 to obtain Ohmic contact with the p-type 4H-SiC piezoresistors. The sheet resistance was measured to be 2.1 M and 26.7 k, corresponding to before and after the annealing process.

After the formation of the 4H-SiC piezoresistors and their Ohmic contact, the 4H-SiC wafer was diced into smaller strips with a dimension of 10 mm × 10 mm. Most importantly, the final fabrication step involved the ultraviolet (UV) scribing etching process to form a square diaphragm on the back of the strip (Fig. 4.10a). In this step, a diode-pumped Nd/YVO$_4$ laser was used with a maximum power of 1.5

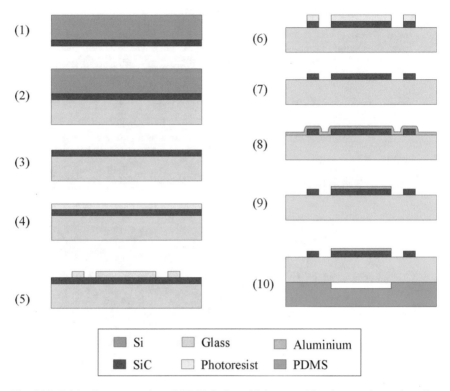

Fig. 4.11 Fabrication process for a MEMS device with integrated heating, sensing and cooling systems. Reprinted with permission from Ref. [64]

kW and an average scribing power of 1–3 W. The 4H-SiC strip with the as-fabricated piezoresistors was positioned in the chamber of the laser system to focus the laser to the strip. After that, the scribing process was performed to etch the back of the 4H-SiC strip. The fabrication process was performed in ablation steps following the layer-by-layer basic until the designed depth of the diaphragm was achieved. The scribing process is much quicker than both conventional wet etching and dry etching processes such as plasma etching and deep reactive-ion etching (DRIE).

Figure 4.10b shows the SEM image of the as-fabricated U-shaped p-type 4H-SiC piezoresistor on top of the pressure sensor. Figure 4.10b illustrates the depth of the diaphragm fabricated by an UV laser scribing process. It is evident that the roughness of the diaphragm is relatively large. Future fabrication techniques should be employed to smooth the surface of the diaphragm formed by UV laser scribing.

4.4.3 Fabrication of MEMS Device with Integrated Cooling System

Figure 4.11 shows multiple fabrication steps for a SiC MEMS device with integrated heating, sensing and microfluidic cooling system [64]. In the first step, highly doped n-type single-crystalline cubic SiC nanofilms were grown on a Si substrate (Fig. 4.11, step 1). Then, the 3C-SiC nanofilms were bonded onto a glass substrate using an anodic bonding method at a pressure of 137 kPa and 1,000 V (Fig. 4.11, step 2).

Next, the Si substrate was removed by wet etching to create a SiC-on-glass platform (Fig. 4.11, step 3). In the next step, a positive photoresist layer was deposited on top of the SiC-on-glass wafer using a spin coater at a speed of approximately 4,000 rpm (Fig. 4.11, step 4). The photoresist was baked at 105 °C for 1.5 min and then patterned to form an array of heating and sensing elements (Fig. 4.11, step 5) employing an ultraviolet light and photoresist developer. The pre-patterned photoresist was then involved in another baking process in 3 min at 120 °C to prepare for the SiC etching process. Next, an inductively coupled plasma (ICP) etching process was employed to pattern SiC heating and sensing components (Fig. 4.11, step 6). In the etching process, the etching chamber pressure was set to about 2 mTorr, followed by an utilisation of a plasma power of approximately 120 W. The etching time was approximately 8 min to completely etch the SiC layer. After the etching step, the photoresist was totally removed (Fig. 4.11, step 7). Then, an aluminium layer was coated on top of the sample employing a sputtering process (Fig. 4.11, step 8). In the next step, electrodes were patterned employing a photolithography process (Fig. 4.11, step 9). Finally, the PDMS channel was then bonded to the backside of the SiC-/glass-assisted by oxygen plasma (Fig. 4.11, step 10).

References

1. R. Yakimova, M. Syväjärvi, M. Tuominen, T. Iakimov, P. Råback, A. Vehanen et al., Seeded sublimation growth of 6H and 4H–SiC crystals. Mater. Sci. Eng., B **61**, 54–57 (1999)
2. J. Jenny, S.G. Müller, A. Powell, V. Tsvetkov, H. Hobgood, R. Glass et al., High-purity semi-insulating 4H-SiC grown by the seeded-sublimation method. J. Electron. Mater. **31**, 366–369 (2002)
3. D. Barrett, R. Seidensticker, W. Gaida, R. Hopkins, W. Choyke, SiC boule growth by sublimation vapor transport. J. Cryst. Growth **109**, 17–23 (1991)
4. H. Li, X. Chen, D. Ni, X. Wu, Factors affecting the graphitization behavior of the powder source during seeded sublimation growth of SiC bulk crystal. J. Cryst. Growth **258**, 100–105 (2003)
5. R. Yakimova, E. Janzén, Current status and advances in the growth of SiC. Diam. Relat. Mater. **9**, 432–438 (2000)
6. R. Puybaret, J. Hankinson, J. Palmer, C. Bouvier, A. Ougazzaden, P.L. Voss et al., Scalable control of graphene growth on 4H-SiC C-face using decomposing silicon nitride masks. J. Phys. D Appl. Phys. **48**, 152001 (2015)
7. T.-K. Nguyen, H.-P. Phan, T. Dinh, T. Toriyama, K. Nakamura, A.R.M. Foisal et al., Isotropic piezoresistance of p-type 4H-SiC in (0001) plane. Appl. Phys. Lett. **113**, 012104 (2018)

8. T.-K. Nguyen, H.-P. Phan, T. Dinh, A. R. M. Foisal, N.-T. Nguyen, D. Dao, High-temperature tolerance of piezoresistive effect in p-4H-SiC for harsh environment sensing. J. Mater. Chem. C (2018)
9. T.-K. Nguyen, H.-P. Phan, T. Dinh, K. M. Dowling, A. R. M. Foisal, D. G. Senesky et al., Highly sensitive 4H-SiC pressure sensor at cryogenic and elevated temperatures. Mater. Des. (2018)
10. A.R. Md Foisal, A. Qamar, H.-P. Phan, T. Dinh, K.-N. Tuan, P. Tanner et al., Pushing the limits of piezoresistive effect by optomechanical coupling in 3C-SiC/Si heterostructure. ACS Appl. Mater. Interfaces. **9**, 39921–39925 (2017)
11. A.R.M. Foisal, T. Dinh, P. Tanner, H.-P. Phan, T.-K. Nguyen, E.W. Streed et al., Photoresponse of a highly-rectifying 3C-SiC/Si heterostructure under UV and visible illuminations. IEEE Electron Device Lett. (2018)
12. A. Qamar, P. Tanner, D.V. Dao, H.-P. Phan, T. Dinh, Electrical properties of p-type 3C-SiC/Si heterojunction diode under mechanical stress. IEEE Electron Device Lett. **35**, 1293–1295 (2014)
13. A. Qamar, H.-P. Phan, J. Han, P. Tanner, T. Dinh, L. Wang et al., The effect of device geometry and crystal orientation on the stress-dependent offset voltage of 3C–SiC (100) four terminal devices. J. Mater. Chem. C **3**, 8804–8809 (2015)
14. A. Qamar, D.V. Dao, J. Han, H.-P. Phan, A. Younis, P. Tanner et al., Pseudo-Hall effect in single crystal 3C-SiC (111) four-terminal devices. J. Mater. Chem. C **3**, 12394–12398 (2015)
15. A. Qamar, H.-P. Phan, T. Dinh, L. Wang, S. Dimitrijev, D.V. Dao, Piezo-Hall effect in single crystal p-type 3C–SiC (100) thin film grown by low pressure chemical vapor deposition. RSC Adv. **6**, 31191–31195 (2016)
16. A. Qamar, D.V. Dao, H.-P. Phan, T. Dinh, S. Dimitrijev, Fundamental piezo-Hall coefficients of single crystal p-type 3C-SiC for arbitrary crystallographic orientation. Appl. Phys. Lett. **109**, 092903 (2016)
17. A. Qamar, D.V. Dao, J.S. Han, A. Iacopi, T. Dinh, H.P. Phan et al., Pseudo-hall effect in single crystal n-type 3C-SiC (100) thin film, in *Key Engineering Materials* (2017), pp. 3–7
18. L. Wang, S. Dimitrijev, J. Han, A. Iacopi, L. Hold, P. Tanner et al., Growth of 3C–SiC on 150-mm Si (100) substrates by alternating supply epitaxy at 1000 C. Thin Solid Films **519**, 6443–6446 (2011)
19. L. Wang, S. Dimitrijev, J. Han, P. Tanner, A. Iacopi, L. Hold, Demonstration of p-type 3C–SiC grown on 150 mm Si (1 0 0) substrates by atomic-layer epitaxy at 1000 °C. J. Cryst. Growth **329**, 67–70 (2011)
20. L. Wang, S. Dimitrijev, A. Fissel, G. Walker, J. Chai, L. Hold et al., Growth mechanism for alternating supply epitaxy: the unique pathway to achieve uniform silicon carbide films on multiple large-diameter silicon substrates. RSC Adv. **6**, 16662–16667 (2016)
21. A. Taylor, J. Drahokoupil, L. Fekete, L. Klimša, J. Kopeček, A. Purkrt et al., Structural, optical and mechanical properties of thin diamond and silicon carbide layers grown by low pressure microwave linear antenna plasma enhanced chemical vapour deposition. Diam. Relat. Mater. **69**, 13–18 (2016)
22. T. Frischmuth, M. Schneider, D. Maurer, T. Grille, U. Schmid, Inductively-coupled plasma-enhanced chemical vapour deposition of hydrogenated amorphous silicon carbide thin films for MEMS. Sens. Actuators, A **247**, 647–655 (2016)
23. M. Lazar, D. Carole, C. Raynaud, G. Ferro, S. Sejil, F. Laariedh et al., Classic and alternative methods of p-type doping 4H-SiC for integrated lateral devices, in *Semiconductor Conference (CAS), 2015 International*, 2015, pp. 145–148
24. Z. Li, X. Ding, F. Li, X. Liu, S. Zhang, H. Long, Enhanced dielectric loss induced by the doping of SiC in thick defective graphitic shells of Ni@ C nanocapsules with ash-free coal as carbon source for broadband microwave absorption. J. Phys. D Appl. Phys. **50**, 445305 (2017)
25. D. Zhuang, J. Edgar, Wet etching of GaN, AlN, and SiC: a review. Mater. Sci. Eng. R: Rep. **48**, 1–46 (2005)
26. S. Pearton, W. Lim, F. Ren, D. Norton, Wet chemical etching of wide bandgap semiconductors-GaN, ZnO and SiC. ECS Trans. **6**, 501–512 (2007)

27. H. Ekinci, V.V. Kuryatkov, D.L. Mauch, J.C. Dickens, S.A. Nikishin, Effect of BCl3 in chlorine-based plasma on etching 4H-SiC for photoconductive semiconductor switch applications. J. Vac. Sci. Technol. B, Nanotechnol. Microelectron.: Mater. Process. Meas. Phenom. **32**, 051205 (2014)

28. P. Yih, V. Saxena, A. Steckl, A review of SiC reactive ion etching in fluorinated plasmas. Phys. Status Solidi B, **202**, 605–642 (1997)

29. L. Jiang, R. Cheung, R. Brown, A. Mount, Inductively coupled plasma etching of SiC in SF 6/O 2 and etch-induced surface chemical bonding modifications. J. Appl. Phys. **93**, 1376–1383 (2003)

30. S. Rysy, H. Sadowski, R. Helbig, Electrochemical etching of silicon carbide. J. Solid State Electrochem. **3**, 437–445 (1999)

31. J. Shor, Electrochemical etching of SiC. EMIS Datarev. Ser **13**, 141–149 (1995)

32. M. Kato, M. Ichimura, E. Arai, P. Ramasamy, Electrochemical etching of 6H-SiC using aqueous KOH solutions with low surface roughness. Jpn. J. Appl. Phys. **42**, 4233 (2003)

33. H. Morisaki, H. Ono, K. Yazawa, Photoelectrochemical properties of single-crystalline n-SiC in aqueous electrolytes. J. Electrochem. Soc. **131**, 2081–2086 (1984)

34. M. Gleria, R. Memming, Charge transfer processes at large band gap semiconductor electrodes: reactions at SiC-electrodes. J. Electroanal. Chem. Interfacial Electrochem. **65**, 163–175 (1975)

35. C. Duval, *Inorganic Thermogravimetric Analysis* (1963)

36. M. Katsuno, N. Ohtani, J. Takahashi, H. Yashiro, M. Kanaya, Mechanism of molten KOH etching of SiC single crystals: comparative study with thermal oxidation. Jpn. J. Appl. Phys. **38**, 4661 (1999)

37. M. Katsuno, N. Ohtani, J. Takahashi, H. Yashiro, M. Kanaya, S. Shinoyama, Etching kinetics of α-SiC single crystals by molten KOH, in *Materials Science Forum* (1998), pp. 837–840

38. L.J. Evans, G.M. Beheim, Deep reactive ion etching (DRIE) of high aspect ratio SiC microstructures using a time-multiplexed etch-passivate process, in *Materials Science Forum* (2006), pp. 1115–1118

39. S. Tanaka, K. Rajanna, T. Abe, M. Esashi, Deep reactive ion etching of silicon carbide. J. Vac. Sci. Technol. B, Nanotechnol. Microelectron.: Mater. Process. Meas. Phenom. **19**, 2173–2176 (2001)

40. P.M. Sarro, Silicon carbide as a new MEMS technology. Sens. Actuators, A **82**, 210–218 (2000)

41. F. Roccaforte, F. La Via, V. Raineri, Ohmic contacts to SiC. Int. J. High Speed Electron. Syst. **15**, 781–820 (2005)

42. Z. Wang, W. Liu, C. Wang, Recent progress in Ohmic contacts to silicon carbide for high-temperature applications. J. Electron. Mater. **45**, 267–284 (2016)

43. J. Riviere, *Solid State Surface Science*, ed. by Green (Marcel Dekker, NY, 1969), p. 179

44. T. Kimoto, J.A. Cooper, *Fundamentals of Silicon Carbide Technology: Growth, Characterization, Devices and Applications* (Wiley, London, 2014)

45. L.M. Porter, R.F. Davis, A critical review of ohmic and rectifying contacts for silicon carbide. Mater. Sci. Eng., B **34**, 83–105 (1995)

46. B. Pécz, G. Radnóczi, S. Cassette, C. Brylinski, C. Arnodo, O. Noblanc, TEM study of Ni and Ni2Si ohmic contacts to SiC. Diam. Relat. Mater. **6**, 1428–1431 (1997)

47. A. Kakanakova-Georgieva, T. Marinova, O. Noblanc, C. Arnodo, S. Cassette, C. Brylinski, Characterization of ohmic and Schottky contacts on SiC. Thin Solid Films **343**, 637–641 (1999)

48. J. Wan, M.A. Capano, M.R. Melloch, Formation of low resistivity ohmic contacts to n-type 3C-SiC. Solid-State Electron. **46**, 1227–1230 (2002)

49. L. Huang, B. Liu, Q. Zhu, S. Chen, M. Gao, F. Qin et al., Low resistance Ti Ohmic contacts to 4H-SiC by reducing barrier heights without high temperature annealing. Appl. Phys. Lett. **100**, 263503 (2012)

50. H. Shimizu, A. Shima, Y. Shimamoto, and N. Iwamuro, Ohmic contact on n-and p-type ion-implanted 4H-SiC with low-temperature metallization process for SiC MOSFETs, Jpn. J. Appl. Phys. **56**, p. 04CR15 (2017)

51. S. Kim, H.-K. Kim, S. Jeong, M.-J. Kang, M.-S. Kang, N.-S. Lee et al, Carrier transport mechanism of Al contacts on n-type 4H-SiC. Mater. Lett. (2018)
52. S. Rao, G. Pangallo, F. Pezzimenti, F.G. Della Corte, High-performance temperature sensor based on 4H-SiC Schottky diodes. IEEE Electron Device Lett. **36**, 720–722 (2015)
53. S. Rao, G. Pangallo, F.G. Della Corte, Highly linear temperature sensor based on 4H-silicon carbide pin diodes. IEEE Electron Device Lett. **36**, 1205–1208 (2015)
54. S. Rao, G. Pangallo, F.G. Della Corte, 4H-SiC pin diode as highly linear temperature sensor. IEEE Trans. Electron Devices **63**, 414–418 (2016)
55. H.P. Phan, T.K. Nguyen, T. Dinh, H. H. Cheng, F. Mu, A. Iacopi et al., Strain effect in highly-doped n-type 3C-SiC-on-glass substrate for mechanical sensors and mobility enhancement. Phys. status solidi A, p. 1800288 (2018)
56. A. Qamar, T. Dinh, M. Jafari, A. Iacopi, S. Dimitrijev, D.V. Dao, A large pseudo-Hall effect in n-type 3C-SiC (1 0 0) and its dependence on crystallographic orientation for stress sensing applications. Mater. Lett. **213**, 11–14 (2018)
57. H.P. Phan, T.K. Nguyen, T. Dinh, A. Iacopi, L. Hold, M.J. Shiddiky et al., Robust free-standing nano-thin SiC membranes enable direct photolithography for MEMS sensing applications. Adv. Eng. Mater. **20**, 1700858 (2018)
58. T. Dinh, H.-P. Phan, T. Kozeki, A. Qamar, T. Namazu, N.-T. Nguyen et al., Thermoresistive properties of p-type 3C–SiC nanoscale thin films for high-temperature MEMS thermal-based sensors. RSC Adv. **5**, 106083–106086 (2015)
59. S.M. Sze, K.K. Ng, *Physics of Semiconductor Devices* (Wiley, London, 2006)
60. S.O. Kasap, *Principles of Electronic Materials and Devices* (McGraw-Hill, New York, 2006)
61. S. Rao, G. Pangallo, F.G. Della Corte, 4H-SiC pin diode as highly linear temperature sensor. IEEE Trans. Electron Devices **63**, 414–418 (2016)
62. G. Brezeanu, F. Draghici, F. Craciunioiu, C. Boianceanu, F. Bernea, F. Udrea et al., 4H-SiC Schottky diodes for temperature sensing applications in harsh environments, in *Materials Science Forum* (2011), pp. 575–578
63. R.S. Okojie, *Fabricating Ultra-thin Silicon Carbide Diaphragms*, Google Patents (2018)
64. T. Dinh, H.-P. Phan, N. Kashaninejad, T.-K. Nguyen, D.V. Dao, N.-T. Nguyen, An on-chip SiC MEMS device with integrated heating, sensing and microfluidic cooling systems. Adv. Mater. Interfaces **1**, 1 (2018)

Chapter 5
Impact of Design and Process on Performance of SiC Thermal Devices

Abstract This chapter describes the influence of the design and fabrication processes on the performance of SiC thermal devices. The chapter first discusses the influence of various substrates on the sensitivity of SiC nanofilms. The impact of doping types and doping levels will be presented. The dependence of the performance of SiC thermal devices on the material morphologies will be mentioned. Finally, this chapter presents the capability of growing SiC nanoscale films and the dependence of the temperature coefficient of resistance on the thickness of SiC films.

Keywords Substrate influence · Doping level · SiC morphologies Thickness influence

5.1 Substrate Influence

Single-crystalline 3C-SiC is typically grown on a Si substrate, creating a 3C-SiC/Si heterostructure. This structure fails to prevent current leakage at high temperatures; hence, it typically requires the transferring of 3C-SiC films to other substrates such as glass. The difference between SiC and substrates in terms of thermal expansion leads to the shift of SiC TCR which is described by the following relationship [1, 2]:

$$\Delta = \frac{-2(\alpha_{1f} - \alpha_{1s})}{1 - \mu_f} \left[\gamma (1 - \mu_f) + \mu_f (1 - \gamma) \right] \tag{5.1}$$

α_{1f} and α_{1s} are the thermal expansion coefficient of the SiC film and substrates, respectively; μ_f and γ are the Poisson's ratio of the SiC thin film and the strain coefficient of resistivity (e.g. gauge), respectively [3–5]. The impact of the substrate is estimated to be less than 100 ppm/K, while the TCR of SiC ranges from 2,000 to above 20,000 ppm/K [6–10]. Therefore, the impact of the substrate is negligible. For 4H-SiC and 6H-SiC, the functional thin films are typically grown on a substrate with the same material (e.g. the same temperature coefficient of expansion) [11–13]. Therefore, the substrate does not affect the temperature sensitivity of these materials.

© The Author(s) 2018
T. Dinh et al., *Thermoelectrical Effect in SiC for High-Temperature MEMS Sensors*, SpringerBriefs in Applied Sciences and Technology, https://doi.org/10.1007/978-981-13-2571-7_5

5.2   Doping Influence

Thermoresistive sensitivity depends upon the activation energy of dopants (e.g. $TCR = -E_a/T^2$). To achieve a high TCR value, the activation energy of the dopants should be large. Nitrogen is typically employed as a donor for n-type SiC materials, while aluminium and boron are used as acceptors for p-type SiC films. Table 5.1 shows the activation energy of these dopants reported in the literature [14–26].

In addition to the doping type, the doping level also contributes significantly to the TCR value of SiC films [11, 25, 27–30]. It is important to note that the activation energy decreases with increasing doping levels. As such, Shor et al. [31, 32] reported that unintentionally doped 3C-SiC has all impurities ionised at temperatures below the room temperature and its resistivity increases with increasing temperatures above the room temperature due to the dominance of the scattering effect. However, at low doping levels (e.g. ~10^{17} cm^{-3}), the temperature at which all impurities are ionised is higher than the room temperature, leading to a decrease in the electrical resistance for the temperature range of the room temperature to around 200 °C. Above this temperature, the resistance increases with increasing temperature, corresponding to a positive temperature coefficient of resistance. For degenerately doped 3C-SiC, impurities are ionised at low temperatures, the scattering effect is dominant, and a constant TCR value of 400 ppm/K was reported [31]. Figure 5.1 illustrates the dependence of the SiC resistivity on doping levels. Recently, Latha et al. [29] has presented the increase in the TCR value in 3C-SiC with decreasing nitrogen doping levels, which is in solid agreement with Shor's work.

The temperature dependence of 6H-SiC resistivity on doping levels was also investigated [33]. For example, at a low doping level (e.g. 10^{17} cm^{-3}), it was assumed that the impurities are ionised at around 0 °C. A high ionisation temperature of 500 °C was observed for a high doping level (e.g. 10^{19} cm^{-3}). At the same high doping level, a higher absolute TCR was measured for the p-type 6H-SiC than the n-type. Figure 5.2 summarises the temperature dependence of resistivity in 6H-SiC with different doping levels [33].

Table 5.1   Activation energy of impurities in SiC in comparison to silicon (Si) [14–26]

Materials	Activation energy		
	Nitrogen	Aluminium (Al)	Boron (B)
Si	0.19	0.069	0.044
3C-SiC	0.04–0.05	0.16	0.73
4H-SiC	0.045–0.125	0.16–0.23	0.25–0.3
6H-SiC	–	–	0.3–0.7

Fig. 5.1 Temperature dependence of resistivity in n-3C-SiC [31]

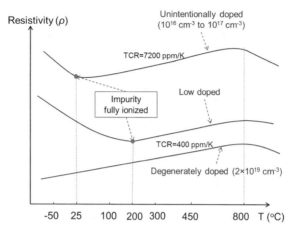

Fig. 5.2 Temperature dependence of resistivity in 6H-SiC [33]

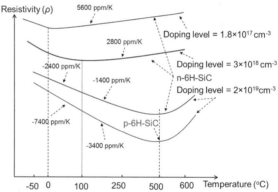

5.3 Morphologies

There are mainly three types of morphologies in material science, including single-crystalline, polycrystalline, and amorphous materials [34]. Single-crystalline materials with an appropriate doping level will offer reasonable conductivity for a wide range of MEMS sensors [35–45]. Since there are no boundaries or defects in the purely single-crystalline structures, the temperature sensing mechanism depends upon the ionisation of impurities or generation of carriers and the scattering effect (Fig. 5.3a). For instance, the impurities in highly doped SiC are ionised at the room temperature, while their mobility decreases with increasing temperature due to the scattering effect [19, 46–49]. This leads to an increase in the electrical resistance with increasing temperature, i.e. a positive temperature coefficient of resistance.

In polycrystalline materials, boundaries and/or defects exist between crystallites, which can trap carriers and create a potential barrier between crystallites [2, 50–52]. This barrier impedes the movement of the carriers and significantly contributes to the electrical resistance of the polycrystalline materials. The resistance of crystallite

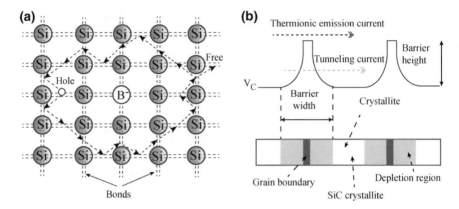

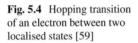

Fig. 5.3 Electrical transport in **a** single-crystalline and **b** polycrystalline silicon semiconductors [8, 54]

Fig. 5.4 Hopping transition of an electron between two localised states [59]

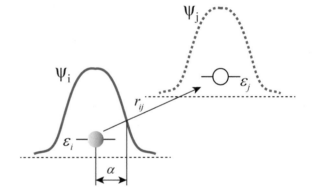

boundaries is sensitive to temperature change. Both thermionic emission current and the tunnelling current increase with increasing temperature, resulting in a significant decrease in the electrical resistance (Fig. 5.3b). At a high level of doping, the barrier height decreases significantly to a small value, leading to a high thermionic emission current and small tunnelling current [49, 51, 53].

In amorphous semiconductors, the charge carrier mobility typically lies between 1 and 10 cm^2/Vs and contributes insignificantly to the electrical transport. The dependence of the electrical conductivity on temperature can be estimated from the density of states (DOS). The tunnelling transitions of charge carriers between localised states in the band tails dominate the transport mechanism of amorphous semiconductors. This is called as a hopping conduction mechanism used to explain the electrical properties of amorphous semiconductors at sufficient high temperatures. Figure 5.4 shows the hopping transition/mechanism of an electron between two localised states i (energy ε_i) and j (energy ε_j) with a distance of r_{ij}. The absorption energy is expressed as follows [6, 55–58]:

Fig. 5.5 Electrical
resistivity of SiC films
depending on deposition
temperatures [60]

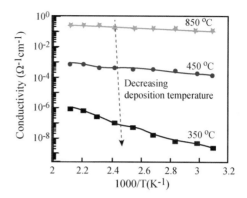

$$v\left(r_{ij}, \varepsilon_i, \varepsilon_j\right) = v_o \exp\left(-\frac{2r_{ij}}{\alpha}\right)\exp\left(-\frac{\varepsilon_j - \varepsilon_i + \left|\varepsilon_j - \varepsilon_i\right|}{2kT}\right) \qquad (5.2)$$

where α and k are the localisation radius and Boltzmann constant, respectively. The conductivity of amorphous semiconductors typically increases with increasing temperatures as presented by Eq. (5.2), showing a large negative TCR value up to $-80,000$ ppm/K [55, 58].

However, due to the high resistivity of amorphous materials, the thermal–electrical effect is typically employed for temperature sensors. For Joule heating-based sensors, an appropriate resistivity is desired, and hence, highly doped polycrystalline or single crystalline is deployed.

5.4 Deposition Temperature

The grain size of SiC crystallites increases with increasing deposition temperatures, resulting in a decrease in the electrical resistivity [5, 50, 61, 62]. Figure 5.5 shows the decrease in the electrical resistivity of SiC films from approximately 10^{-8} to 10^{-1} Ω^{-1} cm^{-1} with increasing deposition temperature from 350 to 850 °C [60]. It is important to note that lower deposition temperatures offer higher resistivity but also more significant change of the SiC electrical resistivity with increasing environmental temperature.

5.5 Geometry and Dimension

The thermoresistive sensitivity of SiC films depends upon their thickness [50, 63]. As such, some studies have proved that a higher TCR is achieved for SiC thinner films [64]. This can be attributed to the free carriers trapped at boundaries of SiC crystal-

Fig. 5.6 Dependence of
temperature sensitivity on
the thickness of the SiC films
[65]

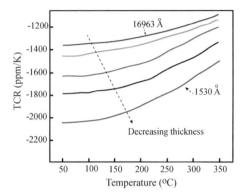

lites. Thinner grown film has a small grain size with more defects and boundaries
between crystallites. Therefore, they have higher resistivity and higher temperature
sensitivity as the defects and boundaries govern the electrical sensing mechanism
of SiC films. The TCR value was measured to be from $-1,400$ to $-2,200$ ppm/K
(Fig. 5.6), corresponding to a film thickness from 1.6963 to 0.153 μm [65].

References

1. P. Hall, The effect of expansion mismatch on temperature coefficient of resistance of thin films. Appl. Phys. Lett. **12**, 212 (1968)
2. F. Warkusz, The size effect and the temperature coefficient of resistance in thin films. J. Phys. D Appl. Phys. **11**, 689 (1978)
3. B. Verma, S. Sharma, Effect of thermal strains on the temperature coefficient of resistance. Thin Solid Films **5**, R44–R46 (1970)
4. F. Warkusz, Electrical and mechanical properties of thin metal films: size effects. Prog. Surf. Sci. **10**, 287–382 (1980)
5. A. Singh, Grain-size dependence of temperature coefficient of resistance of polycrystalline metal films. Proc. IEEE **61**, 1653–1654 (1973)
6. T. Dinh, D.V. Dao, H.-P. Phan, L. Wang, A. Qamar, N.-T. Nguyen et al., Charge transport and activation energy of amorphous silicon carbide thin film on quartz at elevated temperature. Appl. Phys. Express **8**, 061303 (2015)
7. T. Dinh, H.-P. Phan, T. Kozeki, A. Qamar, T. Namazu, N.-T. Nguyen et al., Thermoresistive properties of p-type 3C–SiC nanoscale thin films for high-temperature MEMS thermal-based sensors. RSC Adv. **5**, 106083–106086 (2015)
8. T. Dinh, H.-P. Phan, T. Kozeki, A. Qamar, T. Fujii, T. Namazu et al., High thermosensitivity of silicon nanowires induced by amorphization. Mater. Lett. **177**, 80–84 (2016)
9. H.-P. Phan, T. Dinh, T. Kozeki, A. Qamar, T. Namazu, S. Dimitrijev et al., Piezoresistive effect in p-type 3C-SiC at high temperatures characterized using Joule heating. Sci. Rep. **6** (2016)
10. T. Dinh, H.-P. Phan, T.-K. Nguyen, V. Balakrishnan, L.-H. Cheng, L. Hold et al., Unintentionally doped epitaxial 3C-SiC (111) nanothin film as material for highly sensitive thermal sensors at high temperatures. IEEE Electron Device Lett. **39**, 580–583 (2018)
11. K. Eto, H. Suo, T. Kato, H. Okumura, Growth of P-type 4H–SiC single crystals by physical vapor transport using aluminum and nitrogen co-doping. J. Cryst. Growth **470**, 154–158 (2017)

12. T. Kimoto, A. Itoh, H. Matsunami, Step bunching in chemical vapor deposition of 6H–and 4H–SiC on vicinal SiC (0001) faces. Appl. Phys. Lett. **66**, 3645–3647 (1995)

13. Q. Wahab, A. Ellison, A. Henry, E. Janzén, C. Hallin, J. Di Persio et al., Influence of epitaxial growth and substrate-induced defects on the breakdown of 4H–SiC Schottky diodes. Appl. Phys. Lett. **76**, 2725–2727 (2000)

14. O. Madelung, *Semiconductors—Basic Data* (Springer Science & Business Media, 2012)

15. A.G. Milnes, *Deep Impurities in Semiconductors* (1973)

16. E.M. Conwell, Properties of silicon and germanium. Proc. IRE **40**, 1327–1337 (1952)

17. S. Sze, J. Irvin, Resistivity, mobility and impurity levels in GaAs, Ge, and Si at 300 K. Solid-State Electron. **11**, 599–602 (1968)

18. T. Kimoto, A. Itoh, H. Matsunami, S. Sridhara, L. Clemen, R. Devaty et al., Nitrogen donors and deep levels in high-quality 4H–SiC epilayers grown by chemical vapor deposition. Appl. Phys. Lett. **67**, 2833–2835 (1995)

19. J. Bluet, J. Pernot, J. Camassel, S. Contreras, J. Robert, J. Michaud et al., Activation of aluminum implanted at high doses in 4H–SiC. J. Appl. Phys. **88**, 1971–1977 (2000)

20. Y. Gaoa, S. Soloviev, T. Sudarshan, Investigation of boron diffusion in 6H-SiC. Appl. Phys. Lett. **83** (2003)

21. W. Götz, A. Schöner, G. Pensl, W. Suttrop, W. Choyke, R. Stein et al., Nitrogen donors in 4H-silicon carbide. J. Appl. Phys. **73**, 3332–3338 (1993)

22. W. Hartung, M. Rasp, D. Hofmann, A. Winnacker, Analysis of electronic levels in SiC: V, N, Al powders and crystals using thermally stimulated luminescence. Mater. Sci. Eng., B **61**, 102–106 (1999)

23. J. Pernot, S. Contreras, J. Camassel, J. Robert, W. Zawadzki, E. Neyret et al., Free electron density and mobility in high-quality 4H–SiC. Appl. Phys. Lett. **77**, 4359–4361 (2000)

24. H. Iwata, K.M. Itoh, Donor and acceptor concentration dependence of the electron Hall mobility and the Hall scattering factor in n-type 4H–and 6H–SiC. J. Appl. Phys. **89**, 6228–6234 (2001)

25. P. Wellmann, S. Bushevoy, R. Weingärtner, Evaluation of n-type doping of 4H-SiC and n-/p-type doping of 6H-SiC using absorption measurements. Mater. Sci. Eng., B **80**, 352–356 (2001)

26. H. Matsuura, M. Komeda, S. Kagamihara, H. Iwata, R. Ishihara, T. Hatakeyama et al., Dependence of acceptor levels and hole mobility on acceptor density and temperature in Al-doped p-type 4H-SiC epilayers. J. Appl. Phys. **96**, 2708–2715 (2004)

27. L. Marsal, J. Pallares, X. Correig, A. Orpella, D. Bardés, R. Alcubilla, Analysis of conduction mechanisms in annealed n-Si $1 - x$ C x: H/p-crystalline Si heterojunction diodes for different doping concentrations. J. Appl. Phys. **85**, 1216–1221 (1999)

28. A. Kovalevskii, A. Dolbik, S. Voitekh, Effect of doping on the temperature coefficient of resistance of polysilicon films. Russ. Microlectron. **36**, 153–158 (2007)

29. H. Latha, A. Udayakumar, V.S. Prasad, Effect of Nitrogen Doping on the Electrical Properties of 3C-SiC Thin Films for High-Temperature Sensors Applications. Acta Metall. Sinica (Engl. Lett.) **27**, 168–174 (2014)

30. K. Nishi, A. Ikeda, D. Marui, H. Ikenoue, T. Asano, n-and p-Type Doping of 4H-SiC by wet-chemical laser processing, in *Materials Science Forum* (2014), pp. 645–648

31. J.S. Shor, D. Goldstein, A.D. Kurtz, Characterization of n-type beta-SiC as a piezoresistor. IEEE Trans. Electron Devices **40**, 1093–1099 (1993)

32. J.S. Shor, L. Bemis, A.D. Kurtz, Characterization of monolithic n-type 6H-SiC piezoresistive sensing elements. IEEE Trans. Electron Devices **41**, 661–665 (1994)

33. R.S. Okojie, A.A. Ned, A.D. Kurtz, W.N. Carr, Characterization of highly doped n-and p-type 6H-SiC piezoresistors. IEEE Trans. Electron Devices **45**, 785–790 (1998)

34. H.P. Klug, L.E. Alexander, X-ray diffraction procedures: for polycrystalline and amorphous materials, in *X-Ray Diffraction Procedures: For Polycrystalline and Amorphous Materials*, 2nd edn, ed. by Harold P. Klug, Leroy E. Alexander, (Wiley-VCH, May 1974), p. 992. ISBN 0-471-49369-4

35. C.-M. Ho, Y.-C. Tai, Micro-electro-mechanical-systems (MEMS) and fluid flows. Annu. Rev. Fluid Mech. **30**, 579–612 (1998)

36. M. Mehregany, C.A. Zorman, N. Rajan, C.H. Wu, Silicon carbide MEMS for harsh environments. Proc. IEEE **86**, 1594–1609 (1998)
37. M. Mehregany, C.A. Zorman, SiC MEMS: opportunities and challenges for applications in harsh environments. Thin Solid Films **355**, 518–524 (1999)
38. J.W. Gardner, V.K. Varadan, O.O. Awadelkarim, *Microsensors, MEMS, and Smart Devices*, vol. 1 (Wiley Online Library, 2001)
39. J.W. Judy, Microelectromechanical systems (MEMS): fabrication, design and applications. Smart Mater. Struct. **10**, 1115 (2001)
40. J.W. Gardner, V.K. Varadan, *Microsensors, MEMS and Smart Devices* (Wiley Inc, London, 2001)
41. G.M. Rebeiz, *RF MEMS: Theory, Design, and Technology* (Wiley, 2004)
42. Y. Zhu, H.D. Espinosa, Effect of temperature on capacitive RF MEMS switch performance—a coupled-field analysis. J. Micromech. Microeng. **14**, 1270 (2004)
43. G. Soundararajan, M. Rouhanizadeh, H. Yu, L. DeMaio, E. Kim, T.K. Hsiai, MEMS shear stress sensors for microcirculation. Sens. Actuators, A **118**, 25–32 (2005)
44. A. Koşar, Y. Peles, Thermal-hydraulic performance of MEMS-based pin fin heat sink. J. Heat Transfer **128**, 121–131 (2006)
45. V. Cimalla, J. Pezoldt, O. Ambacher, Group III nitride and SiC based MEMS and NEMS: materials properties, technology and applications. J. Phys. D Appl. Phys. **40**, 6386 (2007)
46. D. Barrett, R. Campbell, Electron mobility measurements in SiC polytypes. J. Appl. Phys. **38**, 53–55 (1967)
47. K. Sasaki, E. Sakuma, S. Misawa, S. Yoshida, S. Gonda, High-temperature electrical properties of 3C-SiC epitaxial layers grown by chemical vapor deposition. Appl. Phys. Lett. **45**, 72–73 (1984)
48. M. Yamanaka, H. Daimon, E. Sakuma, S. Misawa, S. Yoshida, Temperature dependence of electrical properties of n-and p-type 3C-SiC. J. Appl. Phys. **61**, 599–603 (1987)
49. E.A. de Vasconcelos, W.Y. Zhang, H. Uchida, T. Katsube, Potential of high-purity polycrystalline silicon carbide for thermistor applications. Jpn. J. Appl. Phys. **37**, 5078 (1998)
50. A. Singh, Film thickness and grain size diameter dependence on temperature coefficient of resistance of thin metal films. J. Appl. Phys. **45**, 1908–1909 (1974)
51. J.Y. Seto, The electrical properties of polycrystalline silicon films. J. Appl. Phys. **46**, 5247–5254 (1975)
52. J. Pernot, W. Zawadzki, S. Contreras, J. Robert, E. Neyret, L. Di Cioccio, Electrical transport in n-type 4H silicon carbide. J. Appl. Phys. **90**, 1869–1878 (2001)
53. E.A. de Vasconcelos, S. Khan, W. Zhang, H. Uchida, T. Katsube, Highly sensitive thermistors based on high-purity polycrystalline cubic silicon carbide. Sens. Actuators, A **83**, 167–171 (2000)
54. T. Dinh, H.-P. Phan, D.V. Dao, P. Woodfield, A. Qamar, N.-T. Nguyen, Graphite on paper as material for sensitive thermoresistive sensors. J. Mater. Chem. C **3**, 8776–8779 (2015)
55. R. Street, *Hydrogenated Amorphous Silicon* (Cambridge University, Cambridge, 1991)
56. P. Fenz, H. Muller, H. Overhof, P. Thomas, Activated transport in amorphous semiconductors. II. Interpretation of experimental data. J. Phys. C: Solid State Phys. **18**, 3191 (1985)
57. T. Abtew, M. Zhang, D. Drabold, Ab initio estimate of temperature dependence of electrical conductivity in a model amorphous material: Hydrogenated amorphous silicon. Phys. Rev. B **76**, 045212 (2007)
58. M.-L. Zhang, D.A. Drabold, *Temperature Coefficient of Resistivity in Amorphous Semiconductors.* arXiv preprint arXiv:1112.2169, (2011)
59. S. Baranovski, *Charge Transport in Disordered Solids with Applications in Electronics*, vol. 17 (Wiley, 2006)
60. H.S. Jha, P. Agarwal, Effects of substrate temperature on structural and electrical properties of cubic silicon carbide films deposited by hot wire chemical vapor deposition technique. J. Mater. Sci.: Mater. Electron. **26**, 2844–2850 (2015)
61. N.-C. Lu, L. Gerzberg, C.-Y. Lu, J.D. Meindl, A conduction model for semiconductor-grain-boundary-semiconductor barriers in polycrystalline-silicon films. IEEE Trans. Electron Devices **30**, 137–149 (1983)

62. D. Petkovic, D. Mitic, Effects of grain-boundary trapping-state energy distribution on the Fermi level position in thin polysilicon films, in *Proceedings of 20th International Conference on Microelectronics,* 1995, pp. 145–148
63. F. Lacy, Developing a theoretical relationship between electrical resistivity, temperature, and film thickness for conductors. Nanoscale Res. Lett. **6**, 1 (2011)
64. M.I. Lei, *Silicon Carbide High Temperature Thermoelectric Flow Sensor* (Case Western Reserve University, 2011)
65. S. Noh, J. Seo, E. Lee, The fabrication by using surface MEMS of 3C-SiC micro-heaters and RTD sensors and their resultant properties. Trans. Electr. Electron. Mater **10**, 131–134 (2009)

Chapter 6
Applications of Thermoelectrical Effect in SiC

Abstract This chapter describes the applications of the thermoelectrical effect in silicon carbide for a wide range of applications in harsh environments. The thermoresistive effect in a single SiC layer for temperature sensing is known as thermistor, or resistive temperature detector will be mentioned. The temperature sensing in multiple SiC layers will also be discussed. The chapter also presents the application of SiC for thermal sensors based on the Joule heating effect such as thermal flow sensors, convective accelerometers and convective gyroscopes. Other applications towards gas sensing and cooling of MEMS devices are described.

Keywords Temperature sensors · Thermal flow sensors
Convective accelerometers · Convective gyroscopes · Gas sensors
SiC cooling devices

6.1 Temperature Sensors, Temperature Control/Compensation and Thermal Measurement

The thermoresistive sensors have been employed to measure temperature, fluid velocity, acceleration, solar radiation and microwave power. The working principle of thermoresistive sensors is based on the change of sensor resistance due to the variation of surrounding environments (temperature). When a SiC resistive component is applied by a large current, the Joule heating effect will raise the temperature of the component. The resistance change of the SiC component due to the change of surrounding temperature, radiation and fluid velocity will be detected and used as an indicator for measuring the physical parameters. Due to the balance in heat transfer, the following equation presents for thermoresistive sensors placed in the environments [1, 2]:

© The Author(s) 2018
T. Dinh et al., *Thermoelectrical Effect in SiC for High-Temperature
MEMS Sensors*, SpringerBriefs in Applied Sciences and Technology,
https://doi.org/10.1007/978-981-13-2571-7_6

Fig. 6.1 Diagram for
thermoresistive sensors to
measure physical parameters
including flow velocity,
temperature and radiation [2]

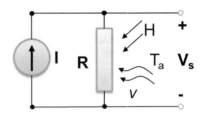

$$\alpha A H + R I^2 = h A(T - T_e) + m_c \frac{dT}{dt} \tag{6.1}$$

where $\alpha A H$ is the radiation power, $R I^2$ is the Joule heating power, and $h A(T - T_e)$ is the dissipated power due to the temperature difference between sensors and the environment. $m_c dT/dt$ is the internal energy variation. α and $A H$ are the transmission coefficient and radiation on the overall area. H, m, c, t are heat transfer coefficient, mass, specific heat and time, respectively. Figure 6.1 shows the diagram for the thermoresistive sensors for detecting various physical parameters, including temperature, flow velocity and radiation.

6.1.1 Thermistors

In industrial and scientific research, temperature sensors, including thermistors, play an important role as a temperature sensing component in the measurement of heat radiation and thermal conductivity of a medium. The utilisation of SiC thermistor has the following benefits [3–12]: (1) high sensitivity (large TCR); (2) miniaturisation (small in size); (3) mechanical stress/strain tolerance; (4) measurement of a wide temperature range to high temperatures (e.g. 600 °C); and (5) a wide range of resistance choices. The main drawback of some SiC thermistors is the nonlinear characteristic. This leads to complex circuitry for the temperature reading. In some cases, the thermistors can be designed in series to adjust the linearity of thermistor. The accuracy of temperature measurement depends upon the following factors [2, 13–15]: (1) measurement circuit; (2) required measurement range; (3) required long-term measurement; and (4) thermistor configuration (e.g. bead, rod and plate). For imprecise temperature measurement, a thermistor can be placed in series with a battery and a resistance metre. For more accurate temperature measurement, SiC thermistors can be used in a Wheatstone bridge configuration/arrangement. To measure the temperature difference, two identical thermistors (e.g. with the same TCR) arranged in two arms of the Wheatstone bridge can be used. Temperature measurement using thermistors can offer a high accuracy of 10^{-4} °C, while the long-term stability of thermistors depends on the range of the temperature change/thermal sock. Conventionally, if long-term stability is the priority, a resistive temperature detector

Table 6.1 Characteristics of SiC temperature sensors in comparison with others [1, 9]

Materials	Sensitivity (ppm/K)	Temperature range (°C)	Response time	Advantages	Disadvantages
Ceramic	-2×10^4 to -6×10^4	-50 to 1000	Fast	Low cost, small size, high sensitivity	Nonlinear response, additional circuitry required, not stable and reproducible
Metal	4×10^3 to -7×10^3	-200 to 650	Slow	High linear response, good long-term stability	Moderate to high cost, low sensitivity, stable and reproducible
Composite	-10^4 to 10^{12}	25 to 100	Slow	Extremely high sensitivity	Moderate to high cost, not stable and reproducible
SiC and semiconductors	-4×10^4 to -10^4	-200 to 1000	Fast	Small size, high sensitivity, compatible with MEMS, integration ability, good long-term stability	Moderate to high cost, additional circuitry required, stable and reproducible

(RTD), such as platinum resistance thermometers, can be used for a wide range of measured temperatures. Alternatively, a thermocouple can be also used as a temperature sensor for a wide range of temperatures. However, thermocouple has less sensitivity for measurement of a small temperature difference. SiC thermistors have been recently demonstrated as highly sensitive temperature sensors for a wide range of temperature measurements. Table 6.1 shows the advantages and disadvantage of various temperature sensors in comparison with SiC temperature sensors [1, 9].

For temperature control applications, the sensor output and signals from the thermistors are processed and employed to control a relay, a valve or operation of transistors. Thermistors can also be used as a sensing element in a temperature alarm system [4, 5, 12, 16–18]. For measurement of thermal conductivity, a thermistor is heated to a high temperature under a high-supply current. The resistance of the thermistor depends on the composition of the medium (e.g. density and pressure) and flow. Therefore, thermistors have been used as gas sensors, pressure sensors and flow sensors. Thermistors can be also used for measurement and control of radiation, electrical quantities (e.g. voltage stabilisers) at audio and radio frequencies, etc [15].

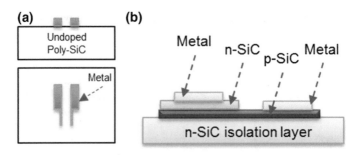

Fig. 6.2 **a** Structure of poly-SiC thermistor [6]. **b** p–n SiC junction temperature sensor [22]

For application of SiC thermistors, the thermoresistive effect in SiC can be employed for temperature sensing, flame detection and control of temperature with high resolution and reliability. Various SiC thermistors have been fabricated using RF-sputtering and chemical vapour deposition (CVD) [4–7, 12]. As such, sputtered SiC films have been demonstrated as highly sensitive temperature sensors for a wide range of temperatures (eg. −100 to 450 °C) [19]. The thermistor equation is typically written as follows:

$$R = R_o \exp\left[B\left(\frac{1}{T} - \frac{1}{T_o}\right)\right] \tag{6.2}$$

where B is known as thermal index presenting the sensitivity of thermistors. The B values of Rf-sputtered polycrystalline SiC thermistors were reported to be 2100 and 3400 for temperature ranges from 0 to 100 °C and 100 to 450 °C, respectively [19]. In another study, a very high sensitivity was achieved with polycrystalline cubic SiC films (eg. $B = 7000$ for 25–200 °C and $B = 5000$ for 200–400 °C) [6]. In addition, a wide range of highly sensitivity temperature sensors can be fabricated using a bi-layer sandwich film of SiC and diamond. Figure 6.2a shows a configuration of SiC thermistors. Table 6.1 summarises various SiC thermistors with the fabrication techniques and thermoresistive sensitivity reported in the literature [20, 21]. While the response time of thermistors has been typically from a few seconds to a few minutes, SiC thermistors were demonstrated with an ultrafast response time of 0.6 s.

6.1.2   p–n Junction Temperature Sensors

p–n junction has been utilised as a temperature sensor since its I–V characteristic changes with temperature change as presented in the following equation [22–27]:

Table 6.2 Temperature sensors using single layer and double layers of SiC [3–8, 12, 16, 25–27, 29, 30]

Configuration	Material	Technique	Temperature range (K)	Sensitivity
Comb resistors	3C-SiC	CVD	300–580	1750–2400 K
Comb resistors	3C-SiC	CVD	300–570	550–4500 K
Rectangular resistor	a-3C-SiC	CVD	300–600	−200 to −16,000 ppm/K
Rectangular resistor	n-3C-SiC	CVD	300–800	200–20,000 ppm/K
Rectangular resistor	p-3C-SiC	CVD	300–600	−2000 to −5200 ppm/K
Zigzag resistors	Diamond/p-Si	Rf-sputtering	200–720	1600–3400
Rectangular resistor	Poly-SiC	Rf-sputtering	275–770	2000–4000
Multi-film layers	SiC/Diamond	CVD	300–670	5000–7000
JFET	6H-SiC	CVD	300–800	–
p–n junctions	4H-SiC	–	300–800	2.2–3.5 mV/K
Schottky diode	Ti/Al/4H-SiC	–	300–600	2.85–5.11 mV/K
PIN diode	4H-SiC	–	300–600	1.63–2.66 mV/K
PIN diode	4H-SiC	–	300–700	2.4–4.5 mV/K

$$I = I_o\left(e^{\frac{qV}{nkT}} - 1\right) \tag{6.3}$$

where I and I_o are the measured currents at temperatures T and T_o, respectively; q is the electron charge; k is the Boltzmann constant; and n is the ideality factor. At forward bias, the voltage of the p–n junction can be defined as $V = 2kT/q \ln(I/I_o)$, and the sensitivity of p–n temperature sensor is $dV/dT = 2k/q \ln(I/I_o)$. Figure 6.2b shows the schematic sketch of a 4H-SiC p–n junction diode which has recently been used as a temperature sensor for a wide range of temperatures from 20 to 600 °C and showed a good sensitivity ranging from 2.2 mV/°C at a forward current density of 0.89 A/cm^2 to 3.5 mV/°C at 0.44 mA/cm^2 [22]. In addition, another study has demonstrated that a 6H-SiC junction gate field-effect transistor (JFET) can be employed as a temperature sensor for a temperature range of 25–500 °C [28]. Table 6.2 summarises the SiC thermistors and other temperature sensors based on the thermoelectrical effect in SiC.

6.2 Thermal Flow Sensors

Thermal flow sensors are utilised to measure the velocity of surrounding fluids and the direction of fluid movement. Thermal flow sensors typically consist of heat-

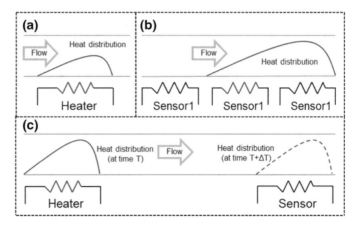

Fig. 6.3 Thermal flow sensors [1, 31]. **a** Hot-film and hot-wire configuration; **b** calorimetric flow sensor; and **c** time-of-flight flow sensor

ing and sensing components. The operation of thermal flow sensors is based on the heat transfer between the heating component and the surrounding environment. Materials with high TCR values and reasonable resistivity are typically used for the sensing component. The main types of thermal flow sensors with the working principle include hot wire/hot film, calorimetric and time of flight, as shown in Fig. 6.3. The operation, fabrication strategy and performance of micromachined thermal flow sensors are summarised in Table 6.3.

6.2.1 Hot-Wire and Hot-Film Flow Sensors

In hot-wire/hot-film flow sensors, the heating and sensing components are integrated into a single element known as heater [49]. The working principle of the hot-wire/hot-film configuration is based on the increase of the heater temperature to a steady state due to the Joule heating effect. When flow is applied, it induces a cooling effect to the heater and hence it changes the heater resistance. This resistance change is typically converted into a voltage change using a Wheatstone bridge. The energy balance between the supply power by the Joule heating effect and the power loss due to the convective cooling effect is as follows [37, 50]:

$$RI^2 = hA(T - T_e) \tag{6.4}$$

where R and h are the electrical resistance of the heater and the heat transfer coefficient, respectively; I is the measured current; and T and T_e are the temperature of the heater and the environment, respectively. The thermoresistive effect in conventional thermal sensing materials including metals and semiconductors has been

Table 6.3 Performance of SiC/glass hot-film air flow sensor in comparison with the literature [32–48]

Heating/sensing material	Substrate material	TCR	Relative resistance change ($\Delta R/R$) in %	Sensitivity
Pt	Alumina	–	22.08	0.029 (s/m)
Polysilicon	Silicon	1100 ppm @ 100 °C	–	5 mV(ms^{-1})–0.5
Pt	Silicon	2490 ppm at @ 25 °C	–	–
CNT	Paper	−750 ppm at @ 25 °C	0.8	<0.0016 (s/m)
Graphite	Paper	–	2.5	0.0062 (s/m)
Graphite	Paper	−2900 ppm at @ 25 °C	3.25	0.0081 (s/m)
3C-SiC	Glass	−20,716 ppm at @ 25 °C	58.36	0.091 (s/m)
Pt	–	–	–	–
Pt	–	3920	–	9.8×10^{-4} mA (m/s)$^{0.5}$/mW or 0.177 mV (m/s)$^{0.5}$/mW
Au	–	–	–	−10 mL/min for N2 and 10 µL/min for water
Au	–	3000		3.06 mV/(ml/min)
Ni	–	–	–	–
Polysilicon		11,000		0.27 V/mL/min
SiC	–	–	–	119 µV/(m/s)/mW
Ge	–	–		232.77 V/W/(m/s)
Ge	–	–		33.33 V/W/(m/s)
Ge	–	−18,000		69.76 V/W/(m/s)
Polysilicon	–	880		1.9 V/W/(m/s)
SiC		1280		0.73 Ω/sccm

typically utilised for developing hot-wire and hot-film flow sensors, because these conventional materials are compatible with micro-/nanomachining technologies and have high TCR values and proper resistivity to function as both heating and sensing elements. For example, thermoresistive effect in platinum was deployed to fabricate hot-film flow sensors [33] with a high sensitivity of 0.177 mV(m/s)$^{-1/2}$/mW and a low power consumption of 45.1 mW.

However, these sensors are not suitable for working in harsh environments including high temperatures and corrosion because of the degradation of the materials. Therefore, SiC has been proven to show the high potential to develop hot-wire/hot-

Fig. 6.4 Experimental set-up for the characterisation of SiC flow sensors. Reprinted with permission from Ref. [51]

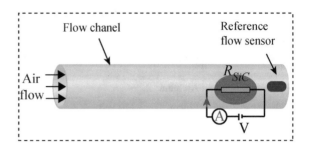

film flow sensors working in harsh environment due to its large band gap and excellent chemical inertness. For example, Lyons et al. reported a hot-wire flow sensor using cubic SiC grown on Si, with a SiC heater and a SiC temperature sensor arranged in two arms of a Wheatstone bridge with two additional external resistors. A fast thermal response time of 3 ms was achieved for the sensor [34]. The SiC sensor has high mechanical strength and good capability to be raised to high temperatures without failures, owing to the high melting temperature of SiC. However, the disadvantage of using SiC on Si for thermal flow sensors is the high power loss due to the high current leakage from SiC to Si at high temperatures. Figure 6.4 shows the schematic sketch of a typical set-up and measurement method for characterisation of SiC flow sensor.

Recently, Balakrishnan et al. [52] have reported MEMS hot-film flow sensors based on a platform of 3C-SiC on a glass substrate. Figure 6.5a and Table 6.4 show the sensor configurations with the dimensions. Figure 6.5b, c shows the sensor output characteristics under different applied air flow velocities in the channel for the small square heater 100 μm × 100 μm and the large square heater 1000 μm × 1000 μm, respectively. The temperature sensor of the small heater configuration has a resistance change of 0.76% while that of the large configuration is 58.36% for the flow range of 0–9 m/s. The inset of Fig. 6.5b, c shows the sensor performance in the laminar flow condition. It is important to note that the heat losses in the laminar regime are relatively low, which increase with the shift of flow conditions to transitional or turbulent.

The sensitivity of the sensor is defined as follows:

$$S = \left(\frac{\Delta R/R}{V_f} \right) = \left(\frac{\frac{R_{flow\_off} - R_{flow\_on}}{R_{flow\_off}}}{V_f} \right) \qquad (6.5)$$

where R_{flow_off} and R_{flow_on} are the initial resistance of the temperature sensor at 0 m/s flow and at the presence of a flow velocity V_f. The sensitivity of the large heater design (1000 μm × 1000 μm) is highest at $S = 0.091$ s/m, while the sensitivity of the small heater design (100 μm × 100 μm) is only approximately 7.32×10^{-4} s/m [52].

Figure 6.6 shows the thermal response time of the SiC flow sensor, a key parameter that represents instantaneous response to the flow rate. Theoretically, the thermal time constant τ was defined from the fitting of the time-dependent electrical resis-

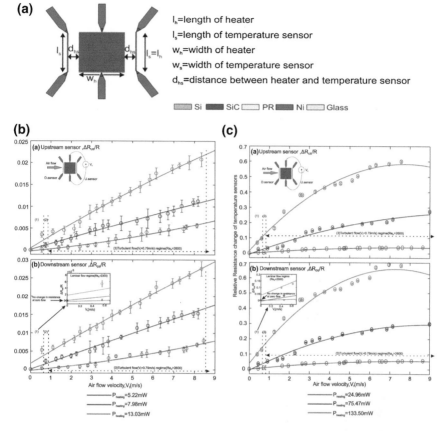

Fig. 6.5 SiC/glass hot-film flow sensors. **a** Sensor configuration; **b** performance of 100 μm × 100 μm SiC thermal flow sensor; and **c** performance of 1000 μm × 1000 μm SiC thermal flow sensor. Reprinted with permission from Ref. [52]

Table 6.4 Sensor dimensions in the design of SiC flow sensor reported by Balakrishnan et al. [52]

Sensor dimensions	1	2	3
Width of the heater w_h (μm)	100	300	1000
Length of the heater l_h (μm)	100	300	1000
Width of the sensor w_s (μm)	10	10	10
Length of the sensor l_s (μm)	100	300	1000
Distance between heater and sensor d_{hs} (μm)	10	10	10

tance $R_s(t) = R_o(1 - e^{-t/\tau})$, where R_o denotes the sensor resistance at zero applied voltage. Experimentally, it is estimated as the time required for the sensor resistance to reach 63% of the steady-state resistance [48]. The heating and cooling phases were

Fig. 6.6 Response of
SiC/glass hot-film flow
sensors. Reprinted with
permission from Ref. [52]

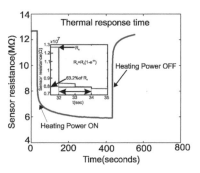

performed (Fig. 6.4) showing a response time of approximately 2 s. The response
time of the SiC/glass flow sensor is comparable with that of other nonreleased thermal
flow sensors constructed on an insulation substrate. This is a much slower response
time in comparison with that of MEMS thermal flow sensors [34, 53, 54] employing
metals, silicon and silicon carbide where the thermal response time is typically less
than 50 ms. The slow response time of the SiC/glass sensor is attributed to the large
thermal mass of the glass substrate. The comparison of the performance of SiC flow
sensors with other thermal flow sensors is presented in Table 6.3.

Recently, unintentionally n-doped SiC has been demonstrated as a hot-film flow
sensor for measuring air flow at elevated temperature. Figure 6.7a shows the resistance change of the sensor with an applied power density of 35 W/cm^2, corresponding
to a temperature of 300 °C. Figure 6.7b illustrates the current response of the SiC
flow sensor to a hot air flow stream (e.g. 130 °C). The measured current decreased
with the applied air flow, indicating that the hot air cooled down the flow sensor
owing to the negative temperature coefficient of resistance of unintentionally doped
SiC film. This leads to an increase of SiC resistance. The high signal-to-noise ratio
was also observed. However, the response time of the flow sensor is relatively slow
(Fig. 6.7b). This is probably due to the power loss to the glass substrate via conduction. To improve the response time of the SiC flow sensors, further work should be
carried out to isolate the SiC material from the substrate. It would be done by etching
of glass to realised SiC-suspended structures.

6.2.2   Calorimetric Flow Sensors

In the **calorimetric** sensing mechanism, the application of flow will induce the
asymmetry of the temperature profile around the heater, leading to the temperature
difference between the upstream and downstream temperature sensors (Fig. 6.3b).
This temperature difference ΔT can be expressed in the following form [31]:

$$\Delta T = T\left(e^{x_1 d_1} - e^{x_2 d_2}\right) \tag{6.6}$$

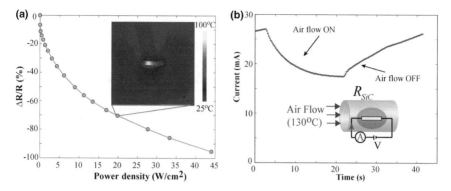

Fig. 6.7 Hot-film flow sensors for monitoring flow at elevated temperatures [51]. **a** The Joule heating effect in the SiC films. The inset presents the infrared image of the SiC heater measured under an application of 20 W/cm². **b** The response of the SiC flow to 130 °C air flow. The inset illustrates the schematic sketch of the experimental set-up. Reprinted with permission from Ref. [51]

where T is the temperature of the heater; d_1 and d_2 are the distances from the heater to the upstream and downstream thermoresistors, respectively; and x_1 and x_2 are two parameters depending on the surrounding fluid characteristics such as velocity, thermal diffusivity and boundary layer thickness. The advantage of this sensing approach is its high sensitivity to small flow rates and ability to sense the flow direction owing to the utilisation of the upstream and downstream thermoresistors. In addition, the development of SiC calorimetric flow sensors is of high interest for harsh environments. Lee et al. reported a design of calorimetric flow sensors with n-type polycrystalline SiC heater surrounded by one downstream and two upstream temperature sensors on silicon nitride/silicon substrates (Fig. 6.8a). This sensor has a sensitivity of 0.73 Ω/sccm, a response time of approximately 15 ms and a negative TCR of -1240 ppm/K for a temperature range of the room temperature to 100 °C [42]. Figure 6.8b shows the change of the resistance of the temperature sensor, corresponding to the change of the flow velocity. It is evident that a higher applied current will lead to a high signal-to-noise response of the thermal flow sensor.

6.2.3 Time-of-Flight Flow Sensors

For time-of-flight flow sensing, it takes a transition time τ for a thermal pulse to travel from a heater to a temperature sensor (Fig. 6.9). The transition time is defined from the flow velocity, thermal conductivity and diffusivity of the medium. However, to date there is no report on the time-of-flight thermal flow sensors employing SiC materials.

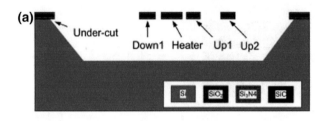

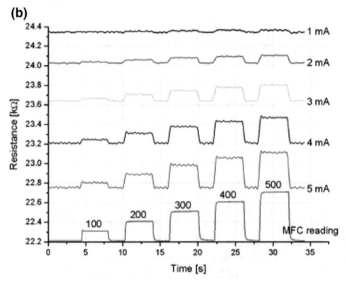

Fig. 6.8 SiC calorimetric flow sensors. **a** Structure of SiC flow sensor with a heater, a downstream temperature sensor and two upstream temperature sensors. **b** Electrical resistance of the upstream sensor under application of different flow rates and different applied currents. Reprinted with permission from Ref. [42, 55]

Fig. 6.9 Working principle of the time-of-flight flow sensor with corresponding circuitry [44, 48]

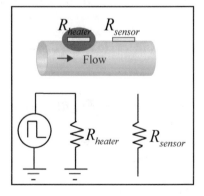

Fig. 6.10 Working principle
of convective accelerometers
[48, 56]

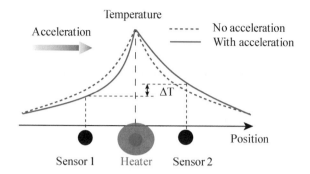

6.3   Convective Accelerometers and Gyroscopes

6.3.1   Convective Accelerometers

Another type of MEMS sensors used for measuring acceleration and angular velocity
is microaccelerometers and gyroscopes. The conventional accelerometers employing
a proof mass sense the acceleration via the piezoresistive effect of designed strain
gauges on the proof mass. The main drawbacks of these mechanical accelerometers
are their low shock resistance and complexity of fabrication. Alternatively, convective accelerometers based on the thermoresistive effect, without using any proof
mass, were introduced to eliminate these disadvantages. The operation of convective accelerometers was first proposed by Weber and then developed using MEMS
technologies [56–58].

The working principle of convective accelerometers is shown in Fig. 6.10. The
Joule heating effect induces the temperature rise around the heater. The inertial
force generated applied accelerations causes the movement of hot fluid bubbles.
This results in an asymmetry of the temperature distribution around the heater. The
temperature difference is sensed by two identical temperature sensors, which are
placed at the same distance from the heater.

Convective accelerometers are required for packaging in a hermetic/enclosed
chamber to prevent disturbance from external environments. These sensors commonly employ a cavity to thermally insulate the heater from the substrate. This
lowers the power consumption of accelerometers.

In the last 20 years, the development of convective accelerometers has been driven
towards high sensitivity and fast response. Table 6.1 summarises the performance
of convective accelerometers reported in the literature. Highly sensitive convective
accelerometers are desired for a wide range of applications. Therefore, the optimisation of the design parameters for the convective accelerometers has been intensively
studied. The impact of design parameters is summarised as follows [59–70]:

(i)   The increase in the supply power and cavity dimensions will improve the sen-
      sitivity.

 (ii)   The frequency response reduces with increasing cavity dimensions.
(iii)   Surrounding gases with low viscosity and high thermal diffusivity will increase
        the sensitivity and frequency response of convective accelerometers.

The development of convective accelerometers is based on single-axis (1D), two-axis (2D) and three-axis (3D) configurations. One-dimensional accelerometers are typically fabricated using conventional micromachining technology [56, 58–62]. A limited number of 2D (planar) convective accelerometers were developed on thin film platforms by standard MEMS technologies [65, 71]. However, it has been facing great challenges for fabrication of 3D convective accelerometers, which can sense the out-of-plane acceleration as the fabrication processes require various complex and non-standard techniques [66]. For example, a monolithic three-axis convective accelerometer has recently been fabricated based on the working principle of 2D accelerometers. Nevertheless, this accelerometer is insensitive to the acceleration applied in z-axis and requires complex circuitry for signal monitoring.

6.3.2 Convective Gyroscopes

Gyroscopes can find a wide range of applications in automotive and other industries, including stabilisation of camera and cell phone, as well as anti-roll fields. Gyroscopes can be used for inertial mouses and displays, and other electronic devices. In terms of working principle, mechanical gyroscopes work based on the Coriolis effect of a proof mass under applied angular velocity.

However, these gyroscopes are fragility, low resistance to vibration and high noise levels. Convective gyroscopes were introduced to address the drawbacks of the conventional mechanical gyroscopes [13, 72–76]. Figure 6.11 shows the working principle of thermal gyroscopes. For example, a dual-axis convective gyroscope was developed with high resolution of 0.05 deg/s, high sensitivity of 0.1 mV/deg/s and bandwidth of 60 Hz at −3 dB [74]. Much effort has paid on development of convective gyroscopes to achieve high sensitivity and capability to detect multi-directional axes. There is no convective gyroscopes developed using SiC.

6.4   Other Applications

6.4.1   Combustible Gas Sensors

The thermoresistive effect in SiC has been employed for chemical composition monitoring such as combustible gases. For instance, in combustion a thermoresistive sensor can sense gases acting on its active surface, owing to the temperature change [18]. Metal oxides and the compounds are typically deployed for combustible gas sensing because of their excellent catalytic properties [77]. There are various combustible

Fig. 6.11 Working principle of thermal gyroscopes [48, 72]

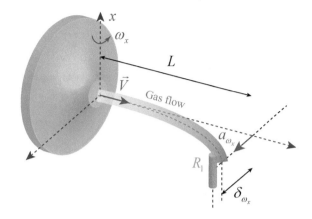

gases in industrial environments, including hydrogen and hydrocarbons. To maintain the safety and efficiency of systems working in the environments, these gases need to be measured. However, the installation of the hazardous gas sensing system is expensive. A solution for this issue is the utilisation of wireless, battery-powered gas sensors would reduce the installation costs, but it requires high-supply power to rise the temperature of the gas sensors up to 500 °C. To minimise the power consumption, these sensors can be miniaturised employing MEMS technologies. SiC has been proven to show as an effective component for combustible gas sensing applications due to its capability to heat up high temperatures with low power consumption.

SiC heaters can function as a gas sensor as shown in Fig. 6.12a. Harley-Trochimczyk et al. [77] reported a simple design of a SiC heater on a low-stress silicon nitride membrane with its capability of being heated to 500 °C at low power consumption of 20 mW (Fig. 6.12b). The SiC was functionalised with platinum nanoparticle-loaded boron nitride aerogel (Pt-BN) as illustrated in Fig. 6.12c.

Figure 6.12d, e, shows the response of the sensor to propane at different humidity and oxygen conditions. The resistance decreases up to 4% when gas concentration increases to 20,000 ppm. The results indicate a high sensitivity and response stability for propane monitoring. A fast response and recovery time of 1 s was reported.

6.4.2 SiC MEMS with Integrated Heating, Sensing and Microfluidic Cooling

The thermoelectrical effect in SiC thin films has been employed to develop a MEMS integrated system, including heaters, temperature sensors and cooling channels. Figure 6.13 shows a recent design of this system, which includes an array of a few heating/sensing SiC units, and a cooling channel is bonded on the back of the chip (Fig. 6.13a). The heater raises the temperature of the device which is sensed by two temperature sensors nearby. When the cooling system is active, the temperature

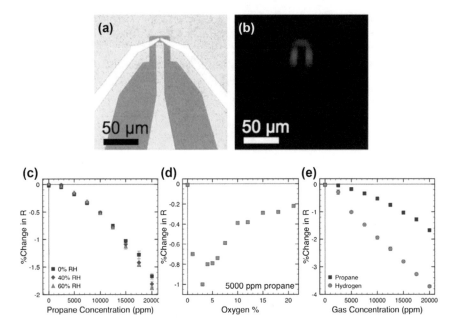

Fig. 6.12 a Optical image of SiC microheater. **b** Microheater with its hot zone. **c** Pt-BN SiC resistance change to propane at 500 °C at different humidity levels. **d** Response of the SiC heater to 5000 ppm propane with different oxygen concentrations. **e** SiC sensor responses to propane and hydrogen. Reprinted with permission from Ref. [77]

of the system decreases, resulting in an increase of the resistance of the temperature sensors. By monitoring this change, the temperature is recorded. Figure 6.13b shows the zoom-in image of a SiC heating/sensing unit. Figure 6.13c–e, h illustrates the as-fabricated components and systems.

The thermal system can be modelled as shown in Fig. 6.14a. The total thermal resistance of the device is the summary of thermal resistance of the substrate R_{glass} and the thermal resistance of the heat sink to the fluid R_{sink}. R_{sink} is calculated as follows:

$$R_{sink} \sim \frac{1}{mC_p\left(1 - e^{-\frac{hA}{mC_p}}\right)} \qquad (6.7)$$

where m and C_p denote the mass flow rate and specific heat capacity, respectively; and h and A are the heat transfer coefficient and the effective area. To improve the thermal cooling efficiency, R_{glass} should be lowered by decreasing the thickness of the substrate. In addition, R_{sink} should be also reduced. In this case, a few parameters can be considered as follows:

* Increasing the fluid velocity
* Increasing the heat transfer area

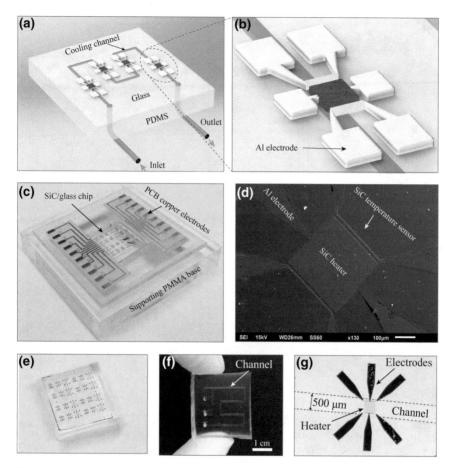

Fig. 6.13 SiC power electronic system with integrated heating, sensing and microfluidic cooling systems. **a** Drawing of a cooling device with a heating/sensing unit (e.g. a SiC heater and two temperature sensors) and a PDMS cooling channel. **b** A zoom-in drawing of a heating/sensing unit. **c** A photograph of the cooling device. **d** SEM image of a SiC heater and two temperature sensors. **e** A photograph of the SiC chip aligned with the channel. **f** A photograph of the PDMS channel. **g** A photograph of the heating/cooling aligned with the cooling channel. Reprinted with permission from Ref. [78]

* Using a fluid with a higher heat transfer coefficient.

Figure 6.14b shows the temperature variation of the MEMS device measured by the SiC temperature sensors. The chip temperature increased from 25 °C to a maximum temperature of 113 °C when a voltage of 2.5 V was applied to the SiC heater (e.g. heating OFF to heating ON). After approximately 150 s, the cooling system was activated with a flow rate of 100 µL/min. This resulted in a temperature decrease of 24 °C, leading to the steady-state temperature of 89 °C. When the cooling

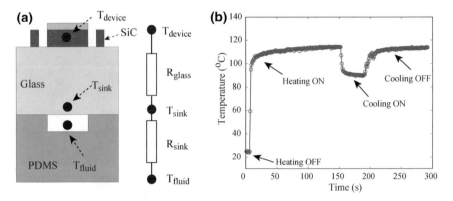

Fig. 6.14 Modelling of the thermal system. **a** The thermal resistances of the system. **b** Response of the system to heating and cooling effect (an applied voltage of 2.5 V and a water flow of 100 μL/min). Reprinted with permission from Ref. [78]

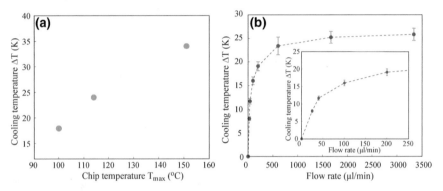

Fig. 6.15 **a** Cooling effect at a constant flow rate of 100 μL/min. **b** Impact of flow rates on the cooling effect (the inset shows the cooling effect in the range of 0–200 μL/min). Reprinted with permission from Ref. [78]

system is OFF (e.g. water flow rate is zero), the temperature of the MEMS device increased to the maximum temperature of 113 °C.

The cooling efficiency of the system is evaluated based on the cooling temperature and chip temperature as follows:

$$\gamma = \frac{\Delta T}{T_{\max} - T_{\text{room}}} \tag{6.8}$$

Figure 6.15a shows the cooling effect of the device for the flow rate of 100 μL/min. The efficiency ranges from 0.24 to 0.28 for different chip temperatures. Table 6.5 shows the performance of the SiC cooling system in comparison with other cooling systems. In addition, the increase of the cooling rate leads to an increase of cooling temperature and a higher efficiency (Fig. 6.15b).

Table 6.5 Comparison of the SiC cooling systems with those reported in the literature

Voltage	Heating	Sensing	Cooling	Production	T_{max} (°C)	ΔT	Efficiency
–	No	No	Yes	Small scale	40	2.9	0.17
–	No	No	Yes	Small scale	50	4.0	0.15
–	No	No	Yes	Small scale	50	6.6	0.264
2 V	Yes	Yes	Yes	Simple, large scale	100	18	0.24
2.5 V	Yes	Yes	Yes	Simple, large scale	113	24	0.273
3 V	Yes	Yes	Yes	Simple, large scale	150	35	0.28

References

1. T. Dinh, H.-P. Phan, A. Qamar, P. Woodfield, N.-T. Nguyen, D.V. Dao, Thermoresistive effect for advanced thermal sensors: fundamentals, design considerations, and applications. J. Microelectromech. Syst. (2017)
2. R.C.S. Freire, S.Y.C. Catunda, B.A. Luciano, Applications of thermoresistive sensors using the electric equivalence principle. IEEE Trans. Instrum. Meas. **58**, 1823–1830 (2009)
3. T. Nagai, K. Yamamoto, I. Kobayashi, Rapid response SiC thin-film thermistor. Rev. Sci. Instrum. **55**, 1163–1165 (1984)
4. T. Nagai, M. Itoh, SiC thin-film thermistors. IEEE Trans. Ind. Appl. **26**, 1139–1143 (1990)
5. E.A. de Vasconcelos, W.Y. Zhang, H. Uchida, T. Katsube, Potential of high-purity polycrystalline silicon carbide for thermistor applications. Jpn. J. Appl. Phys. **37**, 5078 (1998)
6. E.A. de Vasconcelos, S. Khan, W. Zhang, H. Uchida, T. Katsube, Highly sensitive thermistors based on high-purity polycrystalline cubic silicon carbide. Sens. Actuators A **83**, 167–171 (2000)
7. N. Boltovets, V. Kholevchuk, R. Konakova, Y.Y. Kudryk, P. Lytvyn, V. Milenin et al., A silicon carbide thermistor. Semicond. Phys. Quantum Electron. Optoelectron. **9**, 67–70 (2006)
8. C. Chen, Evaluation of resistance–temperature calibration equations for NTC thermistors. Measurement **42**, 1103–1111 (2009)
9. A. Feteira, Negative temperature coefficient resistance (NTCR) ceramic thermistors: an industrial perspective. J. Am. Ceram. Soc. **92**, 967–983 (2009)
10. H. Al-Mumen, F. Rao, L. Dong, W. Li, Design, fabrication, and characterization of graphene thermistor, in *2013 8th IEEE International Conference on Nano/Micro Engineered and Molecular Systems (NEMS)* (2013), pp. 1135–1138
11. C. Yan, J. Wang, P.S. Lee, Stretchable graphene thermistor with tunable thermal index. ACS Nano **9**, 2130–2137 (2015)
12. V. Mitin, V. Kholevchuk, A. Semenov, A. Kozlovskii, N. Boltovets, V. Krivutsa et al., Nanocrystalline SiC film thermistors for cryogenic applications. Rev. Sci. Instrum. **89**, 025004 (2018)
13. H. Chang, X. Gong, S. Wang, P. Zhou, W. Yuan, On improving the performance of a triaxis vortex convective gyroscope through suspended silicon thermistors. IEEE Sens. J. **15**, 946–955 (2015)

14. G.S. Deep, R. Freire, P. Lobo, J.R. Neto, A. Lima, Dynamic response of thermoresistive sensors. IEEE Trans. Instrum. Meas. **41**, 815–819 (1992)
15. M. Prudenziati, A. Taroni, G. Zanarini, Semiconductor sensors: I—Thermoresistive devices. IEEE Trans. Ind. Electron. Control Instrum., 407–414 (1970)
16. P. Fau, J. Bonino, J. Demai, A. Rousset, Thin films of nickel manganese oxide for NTC thermistor applications. Appl. Surf. Sci. **65**, 319–324 (1993)
17. A. Feltz, W. Pölzl, Spinel forming ceramics of the system $Fe_xNi_yMn_{3-x-y}O_4$ for high temperature NTC thermistor applications. J. Eur. Ceram. Soc. **20**, 2353–2366 (2000)
18. Z. Yue, J. Shan, X. Qi, X. Wang, J. Zhou, Z. Gui et al., Synthesis of nanocrystalline manganite powders via a gel auto-combustion process for NTC thermistor applications. Mater. Sci. Eng., B **99**, 217–220 (2003)
19. K. Wasa, T. Tohda, Y. Kasahara, S. Hayakawa, Highly-reliable temperature sensor using rf-sputtered SiC thin film. Rev. Sci. Instrum. **50**, 1084–1088 (1979)
20. E. Obermeier, High temperature microsensors based on polycrystalline diamond thin films, in *The 8th International Conference on Solid-State Sensors and Actuators, 1995 and Eurosensors IX. Transducers' 95* (1995), pp. 178–181
21. M.R. Werner, W.R. Fahrner, Review on materials, microsensors, systems and devices for high-temperature and harsh-environment applications. IEEE Trans. Industr. Electron. **48**, 249–257 (2001)
22. N. Zhang, C.-M. Lin, D.G. Senesky, A.P. Pisano, Temperature sensor based on 4H-silicon carbide pn diode operational from 20 C to 600 C. Appl. Phys. Lett. **104**, 073504 (2014)
23. S.M. Sze, K.K. Ng, *Physics of Semiconductor Devices* (Wiley, New York, 2006)
24. D. Peters, R. Schörner, K.-H. Hölzlein, P. Friedrichs, Planar aluminum-implanted 1400 V 4H silicon carbide pn diodes with low on resistance. Appl. Phys. Lett. **71**, 2996–2997 (1997)
25. S. Rao, G. Pangallo, F. Pezzimenti, F.G. Della Corte, High-performance temperature sensor based on 4H-SiC Schottky diodes. IEEE Electron Device Lett. **36**, 720–722 (2015)
26. S. Rao, G. Pangallo, F.G. Della Corte, Highly linear temperature sensor based on 4H-silicon carbide pin diodes. IEEE Electron Device Lett. **36**, 1205–1208 (2015)
27. G. Chen, S. Bai, A. Liu, L. Wang, R.H. Huang, Y.H. Tao, et al., Fabrication and application of 1.7 kV SiC-Schottky diodes, in *Materials Science Forum* (2015), pp. 579–582
28. J.B. Casady, W.C. Dillard, R.W. Johnson, U. Rao, A hybrid 6H-SiC temperature sensor operational from 25/spl deg/C to 500/spl deg/C. IEEE Trans. Compon. Packag. Manuf. Technol. Part A **19**, 416–422 (1996)
29. S. Rao, G. Pangallo, F.G. Della Corte, 4H-SiC pin diode as highly linear temperature sensor. IEEE Trans. Electron Devices **63**, 414–418 (2016)
30. S.B. Hou, P.E. Hellström, C.M. Zetterling, M. Östling, 4H-SiC PIN diode as high temperature multifunction sensor, in *Materials Science Forum* (2017), pp. 630–633
31. J.T. Kuo, L. Yu, E. Meng, Micromachined thermal flow sensors—a review. Micromachines **3**, 550–573 (2012)
32. S.C. Bailey, G.J. Kunkel, M. Hultmark, M. Vallikivi, J.P. Hill, K.A. Meyer et al., Turbulence measurements using a nanoscale thermal anemometry probe. J. Fluid Mech. **663**, 160–179 (2010)
33. S.-T. Hung, S.-C. Wong, W. Fang, The development and application of microthermal sensors with a mesh-membrane supporting structure. Sens. Actuators, A **84**, 70–75 (2000)
34. C. Lyons, A. Friedberger, W. Welser, G. Muller, G. Krotz, R. Kassing, A high-speed mass flow sensor with heated silicon carbide bridges, in *The Eleventh Annual International Workshop on Micro Electro Mechanical Systems, 1998. MEMS 98. Proceedings* (1998), pp. 356–360
35. A.S. Cubukcu, E. Zernickel, U. Buerklin, G.A. Urban, A 2D thermal flow sensor with sub-mW power consumption. Sens. Actuators, A **163**, 449–456 (2010)
36. R. Ahrens, K. Schlote-Holubek, A micro flow sensor from a polymer for gases and liquids. J. Micromech. Microeng. **19**, 074006 (2009)
37. R.J. Adamec, D.V. Thiel, Self heated thermo-resistive element hot wire anemometer. IEEE Sens. J. **10**, 847–848 (2010)

38. C. Li, P.-M. Wu, J. Han, C.H. Ahn, A flexible polymer tube lab-chip integrated with microsensors for smart microcatheter. Biomed. Microdevice **10**, 671–679 (2008)
39. P. Bruschi, M. Dei, M. Piotto, A low-power 2-D wind sensor based on integrated flow meters. IEEE Sens. J. **9**, 1688–1696 (2009)
40. F. Keplinger, J. Kuntner, A. Jachimowicz, F. Kohl, Sensitive measurement of flow velocity and flow direction using a circular thermistor array, in *GMe Workshop* (2006), pp. 133–137
41. J. Robadey, O. Paul, H. Baltes, Two-dimensional integrated gas flow sensors by CMOS IC technology. J. Micromech. Microeng. **5**, 243 (1995)
42. J.-G. Lee, M.I. Lei, S.-P. Lee, S. Rajgopal, M. Mehregany, Micro flow sensor using polycrystalline silicon carbide. J. Sensor Sci. Technol. **18**, 147–153 (2009)
43. H. Berthet, J. Jundt, J. Durivault, B. Mercier, D. Angelescu, Time-of-flight thermal flowrate sensor for lab-on-chip applications. Lab Chip **11**, 215–223 (2011)
44. E. Meng, P.-Y. Li, Y.-C. Tai, A biocompatible Parylene thermal flow sensing array. Sens. Actuators, A **144**, 18–28 (2008)
45. T. Dinh, H.-P. Phan, D.V. Dao, P. Woodfield, A. Qamar, N.-T. Nguyen, Graphite on paper as material for sensitive thermoresistive sensors. J. Mater. Chem. C **3**, 8776–8779 (2015)
46. T. Dinh, H.-P. Phan, T.-K. Nguyen, A. Qamar, A.R.M. Foisal, T.N. Viet et al., Environment-friendly carbon nanotube based flexible electronics for noninvasive and wearable healthcare. J. Mater. Chem. C **4**, 10061–10068 (2016)
47. T. Dinh, H.-P. Phan, T.-K. Nguyen, A. Qamar, P. Woodfield, Y. Zhu et al., Solvent-free fabrication of biodegradable hot-film flow sensor for noninvasive respiratory monitoring. J. Phys. D Appl. Phys. **50**, 215401 (2017)
48. T. Dinh, H.-P. Phan, A. Qamar, P. Woodfield, N.-T. Nguyen, D.V. Dao, Thermoresistive effect for advanced thermal sensors: Fundamentals, design considerations, and applications. J. Microelectromech. Syst. **26**, 966–986 (2017)
49. S. Noh, J. Seo, E. Lee, The fabrication by using surface MEMS of 3C-SiC micro-heaters and RTD sensors and their resultant properties. Trans. Electr. Electron. Mater **10**, 131–134 (2009)
50. F. Mailly, A. Giani, R. Bonnot, P. Temple-Boyer, F. Pascal-Delannoy, A. Foucaran et al., Anemometer with hot platinum thin film. Sens. Actuators, A **94**, 32–38 (2001)
51. T. Dinh, H.-P. Phan, T.-K. Nguyen, V. Balakrishnan, H.-H. Cheng, L. Hold et al., Unintentionally doped epitaxial 3C-SiC (111) nanothin film as material for highly sensitive thermal sensors at high temperatures. IEEE Electron Device Lett. **39**, 580–583 (2018)
52. V. Balakrishnan, T. Dinh, H.-P. Phan, D.V. Dao, N.-T. Nguyen, Highly sensitive 3C-SiC on glass based thermal flow sensor realized using MEMS technology. Sens. Actuators A Phys. (2018)
53. S. Issa, H. Sturm, W. Lang, Modeling of the response time of thermal flow sensors. Micromachines **2**, 385–393 (2011)
54. C. Sosna, T. Walter, W. Lang, Response time of thermal flow sensors with air as fluid. Sens. Actuators, A **172**, 15–20 (2011)
55. M.I. Lei, Silicon Carbide High Temperature Thermoelectric Flow Sensor (Case Western Reserve University, 2011)
56. A.M. Leung, J. Jones, E. Czyzewska, J. Chen, M. Pascal, Micromachined accelerometer with no proof mass, in *Electron Devices Meeting, 1997. IEDM'97. Technical Digest., International* (1997), pp. 899–902
57. "Accelerometer," ed: Google Patents (1948)
58. A. Leung, J. Jones, E. Czyzewska, J. Chen, B. Woods, Micromachined accelerometer based on convection heat transfer, in *The Eleventh Annual International Workshop on Micro Electro Mechanical Systems, 1998. MEMS 98. Proceedings* (1998), pp. 627–630
59. X. Luo, Y. Yang, F. Zheng, Z. Li, Z. Guo, An optimized micromachined convective accelerometer with no proof mass. J. Micromech. Microeng. **11**, 504 (2001)
60. X. Luo, Z. Li, Z. Guo, Y. Yang, Thermal optimization on micromachined convective accelerometer. Heat Mass Transf. **38**, 705–712 (2002)
61. X. Luo, Z. Li, Z. Guo, Y. Yang, Study on linearity of a micromachined convective accelerometer. Microelectron. Eng. **65**, 87–101 (2003)

62. F. Mailly, A. Giani, A. Martinez, R. Bonnot, P. Temple-Boyer, A. Boyer, Micromachined thermal accelerometer. Sens. Actuators, A **103**, 359–363 (2003)
63. F. Mailly, A. Martinez, A. Giani, F. Pascal-Delannoy, A. Boyer, Design of a micromachined thermal accelerometer: thermal simulation and experimental results. Microelectron. J. **34**, 275–280 (2003)
64. L. Lin, J. Jones, A liquid-filled buoyancy-driven convective micromachined accelerometer. J. Microelectromech. Syst. **14**, 1061–1069 (2005)
65. V.T. Dau, D.V. Dao, S. Sugiyama, A 2-DOF convective micro accelerometer with a low thermal stress sensing element. Based on work presented at *IEEE Sensor 2006: The 5th IEEE Conference on Sensors*, Oct. 22–25, 2006, Daegu, Korea. Smart Mater. Struct. **16**, 2308 (2007)
66. B.T. Tung, D.V. Dao, R. Amarasinghe, N. Wada, H. Tokunaga, S. Sugiyama, Development of a 3-DOF micro accelerometer with wireless readout 電気学会論文誌 E (センサ・マイクロマシン部門誌) **128**, 235–239 (2008)
67. S.-H. Tsang, A.H. Ma, K.S. Karim, A. Parameswaran, A.M. Leung, Monolithically fabricated polymermems 3-axis thermal accelerometers designed for automated wirebonder assembly, in *IEEE 21st International Conference on Micro Electro Mechanical Systems, 2008. MEMS 2008* (2008), pp. 880–883
68. S.-J. Chen, C.-H. Shen, A novel two-axis CMOS accelerometer based on thermal convection. IEEE Trans. Instrum. Meas. **57**, 1572–1577 (2008)
69. U. Park, D. Kim, J. Kim, I.-K. Moon, C.-H. Kim, Development of a complete dual-axis micromachined convective accelerometer with high sensitivity. Sens. IEEE **2008**, 670–673 (2008)
70. J. Bahari, J.D. Jones, A.M. Leung, Sensitivity improvement of micromachined convective accelerometers. J. Microelectromech. Syst. **21**, 646–655 (2012)
71. R. Amarasinghe, D.V. Dao, T. Toriyama, S. Sugiyama, Development of miniaturized 6-axis accelerometer utilizing piezoresistive sensing elements. Sens. Actuators, A **134**, 310–320 (2007)
72. V.T. Dau, D.V. Dao, T. Shiozawa, H. Kumagai, S. Sugiyama, Development of a dual-axis thermal convective gas gyroscope. J. Micromech. Microeng. **16**, 1301 (2006)
73. H. Kumagai, S. Sugiyama, A single-axis thermal convective gas gyroscope. Sens. Mater. **17**, 453–463 (2005)
74. D.V. Dao, V.T. Dau, T. Shiozawa, S. Sugiyama, Development of a dual-axis convective gyroscope with low thermal-induced stress sensing element. J. Microelectromech. Syst. **16**, 950 (2007)
75. V.T. Dau, D.V. Dao, T.X. Dinh, T. Shiozawa, S. Sugiyama, Optimization of PZT diaphragm pump for the convective gyroscope. 電気学会論文誌 E (センサ・マイクロマシン部門誌) **127**, 347–352 (2007)
76. V.T. Dau, D.V. Dao, T. Shiozawa, S. Sugiyama, Simulation and fabrication of a convective gyroscope. IEEE Sens. J. **8**, 1530–1538 (2008)
77. A. Harley-Trochimczyk, A. Rao, H. Long, A. Zettl, C. Carraro, R. Maboudian, Low-power catalytic gas sensing using highly stable silicon carbide microheaters. J. Micromech. Microeng. **27**, 045003 (2017)
78. T. Dinh, H.-P. Phan, N. Kashaninejad, T.-K. Nguyen, D.V. Dao, N.-T. Nguyen, An on-chip SiC MEMS device with integrated heating, sensing and microfluidic cooling systems. Adv. Mater. Interfaces **1**, 1 (2018)

Chapter 7
Future Prospects of SiC Thermoelectrical Sensing Devices

Abstract This chapter presents the future prospect of SiC MEMS thermoelectrical sensing devices in terms of the development of new platform and integration capability of SiC with other materials for high-temperature applications. The chapter also describes the possibility of using the thermoelectrical effect in SiC for possible applications in sensing systems, including resonant sensors. Challenges and opportunities for the development of SiC thermal devices in high-temperature applications will also be discussed.

Keywords SiC on insulator · SiC integrated devices
High-temperature SiC sensors

7.1 Novel Platforms of SiC Films on Insulation Substrates

Single-crystalline 3C-SiC is typically grown on large-area Si substrates at low cost [1–3]. However, Si is not suitable for high-temperature MEMS sensors because of the generation of intrinsic carriers and the degradation at elevated temperatures [4–8]. In addition, the leakage current from 3C-SiC to the Si substrate at elevated temperatures leads to the inaccuracy and unreliability of the SiC electronic devices [9]. Therefore, there has been an increasing demand for the development of new SiC platforms which are can withstand and work reliably at high temperatures [10, 11]. The recent work has demonstrated a new platform of 3C-SiC on a glass substrate [12]. Figure 7.1a shows the structure of 3C-SiC grown on Si with the interface [13]. Figure 7.1b illustrates the corresponding band diaphragm of n-type SiC on Si (left) and p-type SiC on Si (right). The very large band energy gap of SiO_2 completely prevents the movement of electron and hole from 3C-SiC layer to the substrate. The replacement of Si by SiO_2 also eliminates the carrier source from Si. Another advantage of this platform is the excellent transparency of both SiC and glass substrate which is potential for a wide range of applications including transparent electrodes and heaters.

To achieve the single-crystalline 3C-SiC on glass platform, single crystalline is first grown on a Si substrate using low-pressure chemical deposition (LPCVD)

T. Dinh et al., *Thermoelectrical Effect in SiC for High-Temperature MEMS Sensors*, SpringerBriefs in Applied Sciences and Technology,
https://doi.org/10.1007/978-981-13-2571-7_7

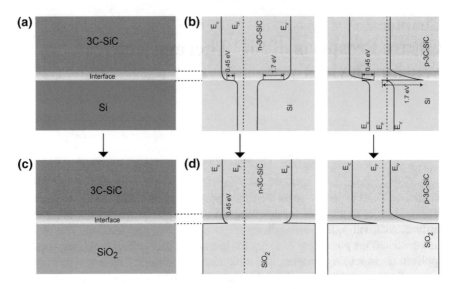

Fig. 7.1 **a** A platform of 3C-SiC on Si with (**b**) the corresponding band energy structures for n-type and p-type 3C-SiC. **c** A new platform of 3C-SiC on an insulator with (**d**) the corresponding band energy diaphragm. Reprinted with permission from Ref. [13]

Fig. 7.2 Formation of a new SiC on glass platform. (1) Growth of SiC nanofilms on Si. (2) Bonding of SiC nanofilms on glass. (3) Formation of SiC/glass platform. Reprinted with permission from Ref. [13]

[14–20]. For small scale of production, SiC films are subsequently released from the Si substrate and then cut into small fragments focused-ion beam (FIB). These fragments are then transferred on to a glass substrate. However, for large-scale production, the SiC on Si platform can be bonded on to a glass wafer/substrate by anodic bonding [21]. The Si substrate is removed by mechanical polishing and wet etching processes. Details on the transferring technique are presented in Fig. 7.2.

Using this platform, a few sensors for high temperatures have been demonstrated including temperature sensors. The temperature sensors for this platform can work in short term with high sensitivity and a high temperature up to 500 °C [22]. Highly sensitive thermal flow sensors based on the hot-film configurations have been also demonstrated [23]. The successful demonstration of this platform for high-temperature sensors has opened a new door for a wide range of electronics working in harsh environments [24].

One main drawback of the current SiC on glass platform is the capability of operating at temperatures less than 500 °C. Another disadvantage is the difficulty of processing or etching of glass substrate which makes great challenges to release SiC structures. Future prospects for the development of SiC electronics could be the transferring of high-quality SiC films on to a silicon-on-insulator (SOI) substrate [13]. The substrate should include a very thin layer of silicon dioxide, and the main substrate layer will be Si. This platform can prevent the leakage current to the substrate, and the fabrication of devices on this platform is simple.

7.2 Integration of SiC Thermoelectrical Devices with Other Materials and Devices

SiC has been well known as a functional material for high-temperature sensing applications [5, 6, 8, 25, 26]. The excellent tolerance of SiC to harsh conditions is obviously an advantage to grow other materials for different applications [26]. The low lattice mismatch between SiC substrate with other piezoelectric materials such as AlN and GaN (Table 7.1) has led to the development of a few platforms for LED and high-frequency applications [27–30]. GaN and AlN can be grown on SiC substrates with low dislocation density and low wafer bow due to the better matching of the thermal expansion coefficient and lattice mismatch. For example, different SiC polytypes (e.g. 3C-SiC, 4H-SiC and 6H-SiC) have been employed to grown GaN and AlGaN for LED applications with low-cost production [31]. In addition, SiC can be employed as a stop-etch barrier for fabrication of GaN and AlN devices due to its excellent chemical inertness [32]. SiC also has a good thermal conductivity; hence, it can naturally reduce the temperature of LED and prolong the lifetime.

A number of applications have been successfully demonstrated using GaN/SiC and AlN/SiC platforms working at high temperatures [33–36]. For example, GaN-on-SiC heterojunction diode has recently employed for temperature sensing in the range of 300–650 K with a high sensitivity of 2.25 mV/K. Figure 7.3 shows the structure of the GaN-on-SiC platform.

Table 7.1 Mismatch between functional GaN and AlN with SiC and other buffer layers [50]

Substrate	Lattice mismatch with GaN (%)	Lattice mismatch with AlN (%)	In-plane thermal expansion coefficient (ppm/K)
Si	17	19	2.6
3C-SiC	3.9	0.95	3.9
GaN	–	2.5	5.6
AlN	2.5	–	4.2

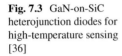

Fig. 7.3 GaN-on-SiC
heterojunction diodes for
high-temperature sensing
[36]

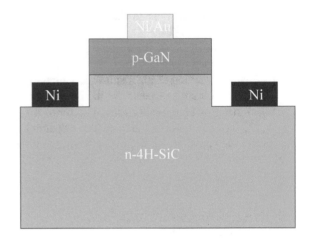

Integration of multifunctional sensing devices on a single power electronic SiC chip is also of high interest. Currently, SiC power devices have been commercialised up to 1.7 kV with operating temperatures up to 300 °C [37–40]. The power SiC devices have a relatively low working temperature due to limitations in packaging and a lack of integrated sensing systems for collecting data at high temperature. Therefore, it is desirable to integrate reliable temperature sensing components on the power electronic chip and wider the operation temperature range of SiC devices. A few studies have demonstrated the SiC temperature sensors with high performance and capability of integration into the SiC power electronics. However, current limitations of these sensors are the linear working range of up to 300 °C, and a large active sensing area or large footprint, leading to the difficulty of miniaturisation. Extensive future studies should be carried out to address this bottleneck.

7.3   SiC Thermal Actuators

Fundamentally, the change of temperature leads to the thermal expansion of SiC structures. Electrothermal actuation of SiC actuators is based on its resistive properties as the Joule heating effect will raise the temperature of the actuator, leading to its expansion [41, 42]. For instance, the differences in thermal expansion of biomaterials (e.g. a metal layer on a SiC layer) have been widely used to bend the structure and create the flexure motion [42–45]. SiC is a suitable material for electrothermal actuators due to its excellent tolerance to high temperatures. However, the high power consumption and long response time limit the applications of thermal actuation.

Currently, a few fundamental SiC structures such as cantilever, bridge and disc have been successfully demonstrated as resonators [43]. In these structures, metal layers (e.g. aluminium, platinum and gold) are deposited on the SiC layers to form the bimorph configuration. When a periodic current is applied, the change of the power distribution on SiC leads to its expansion and subsequently the vibration of microstructures. In addition, the mismatch of the thermal expansion coefficient between the two layers generates bending motion. Employing these fundamental principles, the electrothermal SiC resonators have been demonstrated in the flexure mode with a high resonant frequency of 80 kHz–1 MHz for cantilevers, 171 kHz–1.766 MHz for bridges and 7 MHz for disc [43, 46]. A number of studies reported the SiC bridge structure driven by the electrothermal actuation, thanks to the difference in thermal expansion of the aluminium layer and SiC layer. However, these actuators depend on the metal coated on top of the SiC layer, which could not be used for high-temperature applications as well as lowers the working frequency of such systems.

The SiC electrothermal cantilevers have successfully been employed as filter mixers and temperature up to 100 °C [47]. For the temperature sensing applications, the shift of resonant frequency was used to qualify the temperature variation. Other electrothermal actuation of SiC structures has also successfully fabricated [45]. However, due to high very high working frequency, there has not been any successful experimental demonstration for electrothermal actuation of a single SiC layer. It is expected that advanced fabrication and measurement will take advantage of the high-temperature tolerance of SiC into high-frequency and high quality factor resonators.

7.4 Challenges and Future Developments of SiC Sensing Devices

For three decades, much interest has paid to discover the new SiC technology for high-temperature applications. For example, 6H-SiC and 4H-SiC wafers have been commercialised since 1991 and 1194, respectively (from Cree Research Inc.). However, the cost of these wafers is still very high. Therefore, tremendous progress has been made to improve the quality of SiC thin films, as well as to lower the cost of SiC materials. 3C-SiC is known to be a material choice for MEMS sensors operating in harsh environments due to its capability to be grown on Si substrate at a high quality and low cost. These advancements of growing SiC materials have strongly supported for the development of MEMS sensors towards high-temperature applications.

In the field of high-temperature sensing, SiC has been demonstrated as an excellent candidate for a wide range of electronic devices, owing to its large band gap and high sensitivity. The thermoelectrical effect in SiC has been intensively investigated, and its application for thermal sensing at high temperatures has been proposed. The temperature coefficient of resistance (TCR) of SiC has been measured with a strong dependence on the following aspects [48]:

- Polytypes (e.g. 3C-SiC, 4H-SiC and 6H-SiC);
- Morphologies (e.g. single crystalline, polycrystalline and amorphous structures);

- Doping levels (e.g. a low doping level of $<10^{17}$ cm^{-3} offers a high negative TCR of up to $-20,000$ pmm/K, while a high doping level can turn the TCR to positive);
- Growth conditions such as substrate temperature and thickness of the SiC films.

In addition to the capability of directly measuring the surrounding temperature with high sensitivity, SiC has a suitable resistivity for raising its temperature by the Joule heating effect. The high melting point of SiC is of interest for developing SiC heaters to detect combustible gases such as propane at a high temperature of 500 °C. Local SiC heaters can also be used for the integration on a chip without employing any external heating source. The SiC heaters are also deployed for measurement of fluid flow and are known as thermal flow sensors.

SiC thermal sensors with capability to work in harsh conditions can be developed in future work as SiC has proven to show its high potential for thermal sensing at high temperatures. Currently, a number of power electronics such as MOSFETs have been commercialised, and the difficulty of integrating temperature sensing components has limited the application of these systems at 300 °C. These power electronic devices are expected to be able to work at a higher temperature of up to 600 °C [49]. This requires the significant development of high-performance integrated sensing systems as well as advanced packaging technologies at high temperatures. To achieve this, alternative thermal sensing components as well as new materials with high thermoelectrical sensitivity are needed. In addition, the lack of proper packaging technologies for high temperatures and harsh environments should be addressed to develop thermal sensors, as thermal expansion at high temperatures causes the failures of the devices. The high corrosion in harsh conditions also needs to be taken into account when designing the SiC systems.

Moreover, the development of SiC thermal sensors could be a process of miniaturisation from a microsystem to a nanosystem, which enhances the sensitivity and response of the thermal system. As such, silicon carbide nanowires and nanofilms are expected to be investigated in the near future for integration into nanosystems operating at high temperature and other harsh conditions. In addition, a wide range of measurement without signal saturation is also a desirable factor for SiC thermal sensors. A large working bandwidth with the capability of detecting multiple physical signals is of interest for SiC sensing devices.

Although tremendous progress has been made for the growth of SiC and few power electronics at elevated temperature are commercialised available, the cost of SiC wafers is still very high, especially 4H- and 6H-SiC. 3C-SiC grown on a large area of Si substrates is a suitable choice for lowering the cost of SiC devices. However, the quality of SiC thin films still needs much improvement before employed in the development of electronic devices with commercialised capability. In addition, the influence of growth conditions and fabrication parameters on the thermoelectrical effect of cubic silicon carbide (3C-SiC) thin films should be investigated. These factors are doping levels, thickness of SiC films and type of substrates.

References

1. Q. Wahab, A. Ellison, A. Henry, E. Janzén, C. Hallin, J. Di Persio et al., Influence of epitaxial growth and substrate-induced defects on the breakdown of 4H–SiC Schottky diodes. Appl. Phys. Lett. **76**, 2725–2727 (2000)
2. L. Wang, S. Dimitrijev, J. Han, A. Iacopi, L. Hold, P. Tanner et al., Growth of 3C–SiC on 150-mm Si (100) substrates by alternating supply epitaxy at 1000 °C. Thin Solid Films **519**, 6443–6446 (2011)
3. L. Wang, S. Dimitrijev, J. Han, P. Tanner, A. Iacopi, L. Hold, Demonstration of p-type 3C–SiC grown on 150 mm Si (1 0 0) substrates by atomic-layer epitaxy at 1000 °C. J. Cryst. Growth **329**, 67–70 (2011)
4. J.Y. Seto, The electrical properties of polycrystalline silicon films. J. Appl. Phys. **46**, 5247–5254 (1975)
5. M. Mehregany, C.A. Zorman, N. Rajan, C.H. Wu, Silicon carbide MEMS for harsh environments. Proc. IEEE **86**, 1594–1609 (1998)
6. M. Mehregany, C.A. Zorman, SiC MEMS: opportunities and challenges for applications in harsh environments. Thin Solid Films **355**, 518–524 (1999)
7. M.R. Werner, W.R. Fahrner, Review on materials, microsensors, systems and devices for high-temperature and harsh-environment applications. IEEE Trans. Ind. Electron. **48**, 249–257 (2001)
8. L. Chen, M. Mehregany, A silicon carbide capacitive pressure sensor for high temperature and harsh environment applications, in *Solid-State Sensors, Actuators and Microsystems Conference, 2007. TRANSDUCERS 2007. International* (2007), pp. 2597–2600
9. C. Dezauzier, N. Becourt, G. Arnaud, S. Contreras, J. Ponthenier, J. Camassel et al., Electrical characterization of SiC for high-temperature thermal-sensor applications. Sens. Actuators, A **46**, 71–75 (1995)
10. H.-P. Phan, T. Dinh, T. Kozeki, A. Qamar, T. Namazu, S. Dimitrijev, et al., Piezoresistive effect in p-type 3C-SiC at high temperatures characterized using Joule heating. Sci. Rep. **6** (2016)
11. V. Balakrishnan, T. Dinh, H.-P. Phan, T. Kozeki, T. Namazu, D.V. Dao et al., Steady-state analytical model of suspended p-type 3C–SiC bridges under consideration of Joule heating. J. Micromech. Microeng. **27**, 075008 (2017)
12. A.R.M. Foisal, H.-P. Phan, T. Dinh, T.-K. Nguyen, N.-T. Nguyen, D.V. Dao, A rapid and cost-effective metallization technique for 3C–SiC MEMS using direct wire bonding. RSC Adv. **8**, 15310–15314 (2018)
13. T. Dinh, H.-P. Phan, N. Kashaninejad, T.-K. Nguyen, D.V. Dao, N.-T. Nguyen, An on-chip SiC MEMS device with integrated heating, sensing and microfluidic cooling systems. Adv. Mater. Interfaces **1**, 1 (2018)
14. H.-P. Phan, D.V. Dao, L. Wang, T. Dinh, N.-T. Nguyen, A. Qamar et al., The effect of strain on the electrical conductance of p-type nanocrystalline silicon carbide thin films. J. Mater. Chem. C **3**, 1172–1176 (2015)
15. A. Qamar, H.-P. Phan, J. Han, P. Tanner, T. Dinh, L. Wang et al., The effect of device geometry and crystal orientation on the stress-dependent offset voltage of 3C–SiC (100) four terminal devices. J. Mater. Chem. C **3**, 8804–8809 (2015)
16. A. Qamar, D.V. Dao, J. Han, H.-P. Phan, A. Younis, P. Tanner et al., Pseudo-Hall effect in single crystal 3C-SiC (111) four-terminal devices. J. Mater. Chem. C **3**, 12394–12398 (2015)
17. H.-P. Phan, T. Dinh, T. Kozeki, T.-K. Nguyen, A. Qamar, T. Namazu et al., The piezoresistive effect in top-down fabricated p-type 3C-SiC nanowires. IEEE Electron Device Lett. **37**, 1029–1032 (2016)
18. A. Qamar, H.-P. Phan, T. Dinh, L. Wang, S. Dimitrijev, D.V. Dao, Piezo-Hall effect in single crystal p-type 3C–SiC (100) thin film grown by low pressure chemical vapor deposition. RSC Adv. **6**, 31191–31195 (2016)
19. H.-P. Phan, T. Dinh, T. Kozeki, T.-K. Nguyen, A. Qamar, T. Namazu et al., Nano strain-amplifier: making ultra-sensitive piezoresistance in nanowires possible without the need of quantum and surface charge effects. Appl. Phys. Lett. **109**, 123502 (2016)

20. A. Qamar, D.V. Dao, J.S. Han, A. Iacopi, T. Dinh, H. P. Phan, et al., Pseudo-hall effect in single crystal n-type 3C-SiC (100) thin film, in *Key Engineering Materials* (2017), pp. 3–7
21. H.-P. Phan, H.-H. Cheng, T. Dinh, B. Wood, T.-K. Nguyen, F. Mu et al., Single-crystalline 3C-SiC anodically bonded onto glass: an excellent platform for high-temperature electronics and bioapplications. ACS Appl. Mater. Interfaces **9**, 27365–27371 (2017)
22. T. Dinh, H.-P. Phan, T.-K. Nguyen, V. Balakrishnan, H.-H. Cheng, L. Hold et al., Unintentionally doped epitaxial 3C-SiC (111) nanothin film as material for highly sensitive thermal sensors at high temperatures. IEEE Electron Device Lett. **39**, 580–583 (2018)
23. V. Balakrishnan, T. Dinh, H.-P. Phan, D.V. Dao, N.-T. Nguyen, Highly sensitive 3C-SiC on glass based thermal flow sensor realized using MEMS technology. Sens. Actuators, A (2018)
24. V. Balakrishnan, H.-P. Phan, T. Dinh, D.V. Dao, N.-T. Nguyen, Thermal flow sensors for harsh environments. Sensors **17**, 2061 (2017)
25. D.G. Senesky, B. Jamshidi, K.B. Cheng, A.P. Pisano, Harsh environment silicon carbide sensors for health and performance monitoring of aerospace systems: a review. IEEE Sens. J. **9**, 1472–1478 (2009)
26. T.-K. Nguyen, H.-P. Phan, T. Dinh, A.R.M. Foisal, N.-T. Nguyen, D. Dao, High-temperature tolerance of piezoresistive effect in p-4H-SiC for harsh environment sensing. J. Mater. Chem. C (2018)
27. A. Zubrilov, V. Nikolaev, D. Tsvetkov, V. Dmitriev, K. Irvine, J. Edmond et al., Spontaneous and stimulated emission from photopumped GaN grown on SiC. Appl. Phys. Lett. **67**, 533–535 (1995)
28. E. Kalinina, N. Kuznetsov, V. Dmitriev, K. Irvine, C. Carter, Schottky barriers on n-GaN grown on SiC. J. Electron. Mater. **25**, 831–834 (1996)
29. M.E. Levinshtein, S.L. Rumyantsev, M.S. Shur, *Properties of Advanced Semiconductor Materials: GaN, AlN, InN, BN, SiC, SiGe.* (Wiley, 2001)
30. D. Zhao, S. Xu, M. Xie, S. Tong, H. Yang, Stress and its effect on optical properties of GaN epilayers grown on Si (111), 6H-SiC (0001), and c-plane sapphire. Appl. Phys. Lett. **83**, 677–679 (2003)
31. J. Edmond, A. Abare, M. Bergman, J. Bharathan, K.L. Bunker, D. Emerson et al., High efficiency GaN-based LEDs and lasers on SiC. J. Cryst. Growth **272**, 242–250 (2004)
32. V. Härle, B. Hahn, H.J. Lugauer, S. Bader, G. Brüderl, J. Baur, et al., GaN-based LEDs and lasers on SiC. Phys. Status Solidi (a) **180**, 5–13 (2000)
33. M.A. Khan, X. Hu, A. Tarakji, G. Simin, J. Yang, R. Gaska et al., AlGaN/GaN metal–oxide–semiconductor heterostructure field-effect transistors on SiC substrates. Appl. Phys. Lett. **77**, 1339–1341 (2000)
34. M. Shur, GaN based transistors for high power applications1. Solid-State Electron. **42**, 2131–2138 (1998)
35. S. Madhusoodhanan, S. Koukourinkova, T. White, Z. Chen, Y. Zhao, M.E. Ware, Highly linear temperature sensor using GaN-on-SiC heterojunction diode for harsh environment applications, in *2016 IEEE 4th Workshop on Wide Bandgap Power Devices and Applications (WiPDA)* (2016), pp. 171–175
36. S. Madhusoodhanan, S. Sandoval, Y. Zhao, M. Ware, Z. Chen, A highly linear temperature sensor using GaN-on-SiC heterojunction diode for high power applications. IEEE Electron Device Lett. **38**, 1105–1108 (2017)
37. M. Berthou, P. Godignon, J. Millán, Monolithically integrated temperature sensor in silicon carbide power MOSFETs. IEEE Trans. Power Electron. **29**, 4970–4977 (2014)
38. S. Rao, G. Pangallo, F.G. Della Corte, Highly linear temperature sensor based on 4H-silicon carbide pin diodes. IEEE Electron Device Lett. **36**, 1205–1208 (2015)
39. S. Rao, G. Pangallo, F.G. Della Corte, 4H-SiC pin diode as highly linear temperature sensor. IEEE Trans. Electron Devices **63**, 414–418 (2016)
40. G. Brezeanu, M. Badila, F. Draghici, R. Pascu, G. Pristavu, F. Craciunoiu, et al., High temperature sensors based on silicon carbide (SiC) devices, in *2015 International Semiconductor Conference (CAS)* (2015), pp. 3–10

41. M. Othman, A. Brunnschweiler, Electrothermally excited silicon beam mechanical resonators. Electron. Lett. **23**, 728–730 (1987)
42. E. Mastropaolo, R. Cheung, Electrothermal actuation studies on silicon carbide resonators. J. Vac. Sci. Technol. B Microelectron. Nanometer Struct. Proc. Meas. Phenom. **26**, 2619–2623 (2008)
43. B. Svilicic, E. Mastropaolo, B. Flynn, R. Cheung, Electrothermally actuated and piezoelectrically sensed silicon carbide tunable MEMS resonator. IEEE Electron Device Lett. **33**, 278–280 (2012)
44. E. Mastropaolo, G.S. Wood, I. Gual, P. Parmiter, R. Cheung, Electrothermally actuated silicon carbide tunable MEMS resonators. J. Microelectromech. Syst. **21**, 811–821 (2012)
45. T. Dinh, H.-P. Phan, T. Kozeki, A. Qamar, T. Namazu, Y. Zhu, et al., Design and fabrication of electrothermal SiC nanoresonators for high-resolution nanoparticle sensing, in *2016 IEEE 16th International Conference on Nanotechnology (IEEE-NANO)* (2016), pp. 160–163
46. E. Mastropaolo, I. Gual, R. Cheung, Silicon carbide electrothermal mixer-filters. Electron. Lett. **46**, 62–63 (2010)
47. G. Wood, I. Gual, P. Parmiter, R. Cheung, Temperature stability of electro-thermally and piezoelectrically actuated silicon carbide MEMS resonators. Microelectron. Reliab. **50**, 1977–1983 (2010)
48. T. Dinh, H.-P. Phan, A. Qamar, P. Woodfield, N.-T. Nguyen, D.V. Dao, Thermoresistive effect for advanced thermal sensors: fundamentals, design considerations, and applications. J. Microelectromech. Syst. (2017)
49. N. Zhang, C.-M. Lin, D.G. Senesky, A.P. Pisano, Temperature sensor based on 4H-silicon carbide pn diode operational from 20 °C to 600 °C. Appl. Phys. Lett. **104**, 073504 (2014)
50. Y. Furubayashi, T. Tanehira, A. Yamamoto, K. Yonemori, S. Miyoshi, S.-I. Kuroki, Peltier effect of silicon for cooling 4H-SiC-based power devices. ECS Trans. **80**, 77–85 (2017)

Printed in the United States
By Bookmasters